Erwin Dee Kord (Ed.)

Outtrim Railway Line

Erwin Dee Kord (Ed.)

Outtrim Railway Line

Victoria (Australia), South Gippsland Railway Line

Solv

Imprint

Publisher:
Solv is a trademark of
International Book Market Service Ltd., 17 Rue Meldrum, Beau Bassin, 1713-01 Mauritius
Email: info@bookmarketservice.com
Website: www.bookmarketservice.com

Published in 2011

Printed in: U.S.A., U.K., Germany. This book was not produced in Mauritius.

ISBN: 978-613-9-05791-7

Contents

Articles

References

Outtrim_railway_line

Outtrim railway line, Victoria	
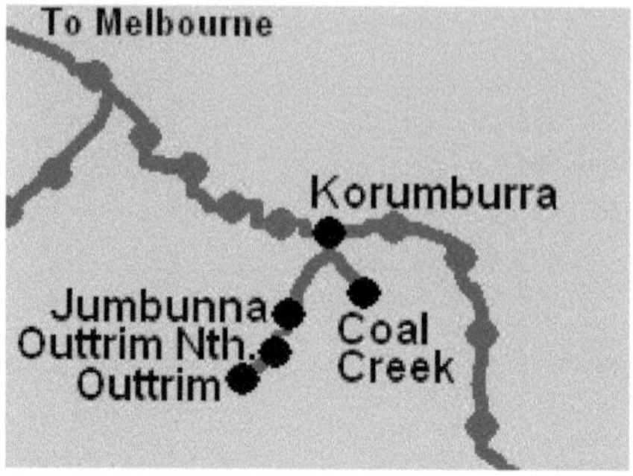	
Line details	
Commenced	6 March 1885
Opened	1894 to Jumbunna 1896 to Outtrim
Completed	1896
Closed	Closed in 2 stages 4 September 1951 Outtrim to Jumbunna, 1 October 1953 Jumbunna to Korumburra
Fate	Closed
Stations	3
Tracks	Single
Former connections	Joined South Gippsland line at Korumburra, Coal Creek line branched off.
Rail transport in Victoria	
List of stations **Closed railways**	

The **Outtrim railway line** was a steam-era regional railway line in Victoria, Australia. It was also part of the former *Great Southern Railway* (South Gippsland line), branching from the current Pakenham line at Dandenong, extending out into the South Gippsland region, The Outtrim section of the line was opened in the early 1890s and branched off the main line just after Korumburra railway station.

History

The Outtrim Line was part of the former *Great Southern Railway*, South Gippsland line. The line was well known for its sharp curves and spectacular scenery, it was also one of the last lines to offer a 'Mixed Passenger and Goods' service in Victoria. This service continued to Yarram until the early 1980s.

The first sod of soil turned in the construction of the Outtrim railway was on 6 March 1885. The line was then opened from Korumburra to Jumbunna in 1894 and the final section to Outtrim in 1896, it was developed as a branch off the main line at Korumburra Station(Korumburra). Its main purpose was to serve the coal mining areas around Outtrim. Shortly after line branched off at Korumburra it had a second branch going to Coal Creek. The Outtrim line was eventually closed in the early 1950s in two stages, on 4 September 1951 from Jumbunna to Outtrim and on 1 October 1953 from Jumbunna to Korumburra. The Outtrim line was closed at a time when Victorian Railways were closing many other short branch lines throughout Victoria.

The Jumbunna railway embankment[1] and railway line viaduct[2] are listed in the Victorian Heritage Inventory.

See also

- List of closed regional railway stations in Victoria
- Transportation in Australia

References

[1] "Jumbunna Railway Embankment, [[Victorian Heritage Inventory (http://vhd.heritage.vic.gov.au/vhd/heritagevic#detail_places;43999)] Number H8021-0042"]. *Victorian Heritage Database*. Heritage Victoria. .

[2] "Jumbunna Railway Line Viaduct, [[Victorian Heritage Inventory (http://vhd.heritage.vic.gov.au/vhd/heritagevic#detail_places;44001)] Number H8021-0041"]. *Victorian Heritage Database*. Heritage Victoria. .

Outtrim,_Victoria

Outtrim Victoria	
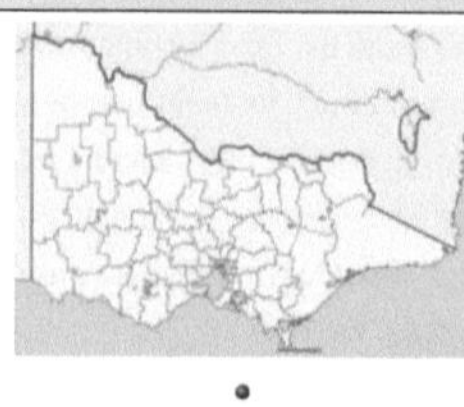 Outtrim	
Postcode:	3951
Location:	151 km (94 mi) SE of Melbourne
LGA:	South Gippsland Shire
State District:	Bass
Federal Division:	McMillan

Outtrim is a locality in Victoria, Australia. It is located south of Korumburra off the Korumburra to Wonthaggi Road and is southeast of Melbourne.

History

The area was originally used for farming in the mid 1800s, but coal was discovered in the area in 1892 south of Outtrim Railway Station. Outtrim and Jumbunna coal fields developed in the mid 1890s, with access to the markets for the coal being established via the Outtrim railway line.[1]

The township grew up by the mine and was named after the Honourable Albert Richard Outtrim, MLA, Minister of Mines in the Victorian Government at the time. By the 1940s only a small number of companies remained working the mines in the area. By the 1980s the township of Outtrim had virtually disappeared with only a few buildings remaining in the town area, and the land returning to farming.

A school was built and opened in 1895. When the coal mines were at their peak more than 200 children attended the four-roomed school. When coal production declined and then ceased with the closing of the mines in 1914, the population dwindled very quickly with most moving to Wonthaggi where a large mine opened in 1909.[2]

The land around Outtrim is now largely cleared of native vegetation and is used for agriculture, including dairying.

Cemetery

Outtrim Cemetery, established circa 1894, is set in woodland one kilometre east of Outtrim Recreation Reserve on the Leongatha South Road. The cemetery has fifteen marked graves, with the earliest dating from 1902. It is believed that there are a total of 220 burials in the cemetery, with last burial being in 1946. The cemetery is listed on the Victorian Heritage Inventory.[3]

The entire cemetery reserve is now reserved permanently for nature conservation purposes and is managed by Parks Victoria.

References

[1] "Outtrim Coal Mine, [[Victorian Heritage Inventory (http://vhd.heritage.vic.gov.au/vhd/heritagevic#detail_places;43996)] Number H8021-0045"]. *Victorian Heritage Database*. Heritage Victoria. .

[2] Outtrim Cemetery (http://www.australiancemeteries.com/vic/sthgippsland/outtrim.htm), Australian Cemeteries

[3] "Outtrim Cemetery, [[Victorian Heritage Inventory (http://vhd.heritage.vic.gov.au/vhd/heritagevic#detail_places;44895)] Number H8020-0023"]. *Victorian Heritage Database*. Heritage Victoria. .

South_Gippsland_railway_line

South Gippsland railway line, Victoria	
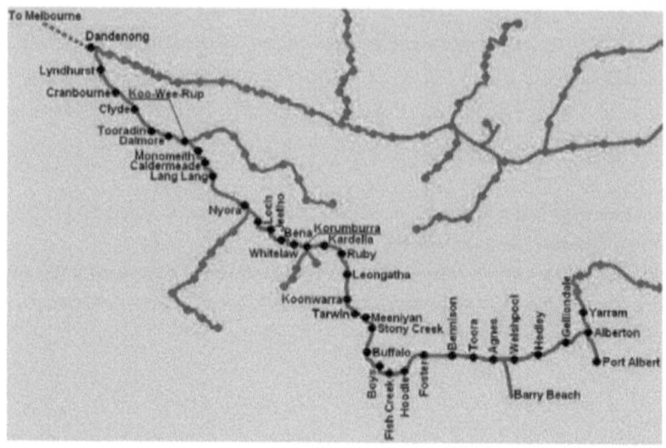	
Line details	
Commenced	1888
Completed	1892
Closed	14 December 1994
Fate	Suburban Cranbourne line, South Gippsland Tourist Railway, Great Southern Rail Trail
Length	126.7 km
Stations	18
Tracks	Single
Used by	South Gippsland Tourist Railway (Nyora - Leongatha), Suburban Cranbourne line (Dandenong - Cranbourne)
Connections	Gippsland line
Former connections	Wonthaggi line, Barry Beach line, Woodside line, Outtrim line, Strzelecki Line
Rail transport in Victoria	
List of stations **Closed railways**	

The **South Gippsland railway line** opened in 1892, branching from the Gippsland line at Dandenong, and extending to Port Albert. Much of it (the section up to Leongatha) remained in use until July 1993 (the track as far as Leongatha is still intact), however only a small portion is in use today, as the suburban Cranbourne line.

In addition to the Cranbourne line, the section from Nyora to Leongatha is used by the South Gippsland Tourist Railway, and Leongatha to Foster has been converted to the Great Southern Rail Trail.

History

The Melbourne and Suburban Railway Company opened a line from Princes Bridge railway station to Punt Road (Richmond) and South Yarra in 1859 and extended to Dandenong in 1879. The **South Gippsland railway line** (also known as the **Great Southern Railway**) was opened from Dandenong to Cranbourne in 1888 and extended to Koo Wee Rup, Nyora and Loch in 1890, Korumburra and Leongatha in 1891 and Welshpool, Alberton and Port Albert in 1892. The section from Alberton to Port Albert was closed in the 1940s. A branch line was built from Alberton to Yarram and Woodside in 1921.

The line was well known for its sharp curves and spectacular scenery, and was also one of the last lines to offer a 'Mixed Passenger and Goods' service in Victoria. The section from Yarram to Woodside was closed in the early 1970s, with the section from Barry Beach Junction and Welshpool to Yarram closed in 1987 when the mixed goods service closed. From this point until about 1992, the track beyond Agnes received minimal usage, although a short branch leading from Agnes to Barry Beach was used extensively for goods traffic to serve the oil platforms in Bass Strait.

The Barry Beach service ceased in 1992 and with it all usage of the train line beyond Leongatha. This section of track was then dismantled in 1994, requiring the strengthening of the large Tarwin bridge near Meeniyan, next to the South Gippsland Highway, to allow the track machine onto the track to do the dismantling. On Saturday 24 July 1993 the last V/Line passenger train ran to Leongatha. P class diesel locomotive P18 hauled the passenger train from Melbourne to Leongatha and back with a FSH 4 carriage set. The section from Nyora to Leongatha was transferred to the South Gippsland Railway in 1994 and trains have not operated beyond Cranbourne since January 1998.[1]

Branch lines

A 5 km-long narrow-gauge horse-drawn tramway was built from Welshpool to Welshpool Jetty in 1905. It was closed in 1941.

The mountainous Strzelecki branch line was built from Koo Wee Rup to Strzelecki which opened in 1922. This line was closed in stages throughout the 1930s, '40s and '50s.

The Wonthaggi branch line was built from Nyora to Wonthaggi opening in 1910 to serve the State Coal Mine, a small extension to the line was opened in 1917. The Wonthaggi line was closed in 1978.

The Outtrim branch line was built from Korumburra to the coal mine at Outtrim in two stages, the initial stage to Jumbunna was opening in 1894 with the final stage to Outtrim opening in 1896, The line was closed in two stages the first from Jumbunna to Outtrim on 4 September 1951 and then from Jumbunna to Korumburra on 1 October 1953.

Services

Passenger services operated on the line since opening. Services from Melbourne to Leongatha and Yarram were withdrawn on 6 June 1981, with replacement buses starting three days later.[2] The Yarram train was a loco hauled mixed train, while the Leongatha train was a DRC railcar hauling MTH carriages. To fill the gap local trains were introduced on a three month trial from Dandenong to Lang Lang, being withdrawn on 3 October 1981 due to insufficient patronage. Services to Leongatha were restored on 9 December 1984 but were again withdrawn on 24 July 1993.[2]

The line became unused beyond Cranbourne with the withdrawal of freight services from Koala siding in January 1998. The exception was a tourist operation which would become known as South Gippsland Railway which commenced operation between Nyora and Leongatha. In 1995 the section between Dandenong and Cranbourne was electrified and a station added at Merinda Park, as part of a $27 million Federal Government funded project.[3]

The Barry Beach freight service ceased in 1992 and with it all usage of the train line beyond Leongatha. V/Line passenger services ceased to Leongatha on 24 July 1993 with last train to Melbourne. By the mid 1990s only T, Y and P class diesel locomotives were used on the line from this time due to their low axle loads, with a 15 km/h speed limit applying to parts of the track.[1] This continued until the January 16, 1998, when the Koala Siding (near Nyora) to Spotswood sand train ceased operation.[1] It took a normal train 2 hours and 20 minutes to travel from Spencer Street Station to Leongatha, with the current bus service completing the journey in a similar time.

The South Gippsland Tourist Railway at Korrumburra loaned out K 190 during the summer of 1995-1996, repainted the engine and tender a more noticeable green. She was once again returned for service with the S.G.T.R during the summer of 1996-1997.

Steamrail, railfan enthusiasts trains

A number of special railway enthusiast services were operated along the line in the mid 1980s to late 1990s, because the line was near closure. The last railfan steam special to run beyond Leongatha was on 13 November 1988, when steam locomotives D3.639 and K153 ran to Foster. The train ran to Foster due to there being a turntable for usable condition. The line between Foster and Barry Beach had minimal usage and the line beyond Welshpool to Yarram was closed and not in usable condition and Toora was the only other main settlement beyond Foster that a steamer could have run to. There were a few diesel hauled heritage tours and special trains that ran beyond Leongatha around the late 1980s to early 1990s.

K153 ran the last steam train to Yarram on 26 October 1987. It also hauled a trip a month earlier to Welshpool on 6 September 1987. There were numerous steam hauled tours to Leongatha in the early 1990s. In 1993 there were two notable trips to Leongatha that were steam hauled. The first one was on 5 March, which was to be a K triple header to Korumburra and Leongatha. One of the locomotives scheduled to haul this train, K190, broke down that morning, so locomotives K153 and K183 made hard work of hauling the long consist of wooden carriages. On 25 July 1993, the apparent last steam train to South Gippsland, ran with K190 and K183 hauling a long consist of wooden carriages. The train ran shuttles between Korumburra and Leongatha.

After the withdrawal of passenger services to Leongatha, a few more steam hauled trips were organised before the remainder of the line became booked out between Cranbourne and Nyora. In April 1994, K190 & K183 again ran a tour to Leongatha, but with a much smaller consist this time. J515 ran to Leongatha on 14 October 1995, which was organised by the Seymour Heritage Railway & ARHS. It was the first time for seven year that a J ran on South Gippsland line. D3.639 and K183 ran to Korumburra and Leongatha on 25 September 1997, which was the last Steamrail train to ever traverse the line. During this time D3.639 was on loan to SGTR

The last steam train on the mainline before closure of the line was steam locomotive K190 operated by Steamrail Victoria. It returned in April 1998 after it was leased to the South Gippsland Tourist Railway (SGTR) in December 1996. SGTR had loaned the loco during the summer of 1995-1996. The locomotive again returned for the summer of 1996-1997. K190 returned on December 7, 1996, when Steamrail ran a tour to Korumburra and Leongatha. Diesel loco T345 was sent down to SGR as well. K183 double headed with K190 to Nyora with T345 dead attached. K190 & T345 were detached at Nyora and then K183 ran solo to Korumburra and Leongatha. K190 followed with the SGR air conditioned cars before running a shuttle to Leongatha. T342, T345 and DERM 55 ran a rare Saturday tourist rail service K183 the Steamrail train back to Melbourne while K190 stayed down at SGTR for the summer.

Current status

The track beyond Leongatha to Yarram and the Barry Beach branch line was dismantled in 1994, but removal of other infrastructure along the line such as level crossing signals took place around mid to late 1992, soon after the last train ran to Barry Beach. The section between Leongatha - Foster was turned into the Great Southern Rail Trail in 1998. The section between Foster and Yarram is planned to be a future extension of the trail.

The branch line from Nyora to Wonthaggi was closed in 1978 and later dismantled, now forming the Bass Coast Rail Trail. A significant section of the former Wonthaggi Line has been leased or sold to farmers, while other sections have been cut away to widen highways and main roads in the coastal South Gippsland region.

The section from Dandenong to Cranbourne was electrified in 1995 and is now part of the suburban rail network as the Cranbourne Line. The first level crossing on the closed section of the line, the South Gippsland Highway crossing in Cranbourne has since been paved over. The Victorian Transport Plan of 2009 has stated that the Cranbourne line will be extended 2 kilometres to a new station at Cranbourne East by 2015. The new Cranbourne East station will be built near Renyard Street and the Casey Complex.[4]

The line from Nyora to Leongatha is now used by the South Gippsland Tourist Railway. However some bridges have been repaired between Leongatha and Korumburra which has allowed the South Gippsland Tourist Railway to restore its services on this section of the track after it was prevented from using it due to the declining condition of the bridges.

The remaining section of track from Cranbourne to Nyora is disused but mostly still intact. Two metres of track was pulled up at Koo Wee Rup to put toilets at the bus and coach stop. The Wonthaggi desalination plant pipeline has also made it necessary to remove three more sections of the line. The first section is about 100 metres on the down side of Monomeith Road where about 50 metres of track have been removed. The second is about 5% m on the up side of Caldermeade Road where about thirty metres of track have been removed and the third, is about halfway between Caldermeade and Lang Lang stations. At this time the track from Cranbourne East to Nyora has not yet been lifted to be converted into the proposed rail trail.

Much of the operational and safeworking infrastructure remains in place in this section, including signalling equipment, level crossings, and easements not having been otherwise leased or sold. Station platforms are also in place, but mostly without station buildings, such as Lang Lang, Koo-Wee-Rup, and Tooradin. Parts of the track have warped due to erosion or movement of the supporting groundwork and displacement of sleepers, these are especially evident near Lang Lang, Tooradin and north of Koo-Wee-Rup. Other potential problems include questions on safety of certain bridges and culverts along the route.

Prior to the 1999 Victorian election, the state Labor Party made a promise to return passenger services to Leongatha. In 2008 a report commissioned by the Victorian Department of Transport found the cost of returning passenger services to Leongatha to be unjustifiably high, at an estimated $72 million, and that only 20 per cent of respondents surveyed about their transport needs considered restoring train services to be the main priority. Instead $14.7 million will be spent on road coach service upgrades.[5] [6]

The former Minister for Public Transport Lynne Kosky has stated that the State Government is providing funding for development of a rail trail between Cranbourne East and Nyora to support tourism in South Gippsland. Such a move may result in the track being lifted, thus ending any hope that the line may reopen in the future. However it still could be possible for the line to reopen in the future even the new rail trail does go ahead between Cranbourne East and Nyora due to reports of a proposed new route for the South Gippsland line if it ever reopens. A proposal is to relocate the tracks of the South Gippsland line prior to Koo-Wee-Rup. The proposed new location of the line would branch off from Pakenham instead of sharing the line between Dandenong and Cranbourne, locate around Pakenham-Koo-Wee- Rup Road then join the existing tracks at Koo-Wee-Rup. This would mean the track could be dismantled between Cranbourne East and Koo-Wee-Rup via Clyde, Tooradin and Dalmore for a new rail trail but not from Koo-Wee-Rup to Nyora unless it locates alongside the actual rail tracks. Another proposal is to have the line branched off from Drouin and diverting through Poowong and joining the existing tracks at Nyora. This would mean that the tracks could be dismantled between Koo-Wee-Rup and Nyora to allow a rail trail to be constructed all the way from Cranbourne East.

References

[1] "Farewell — The Sand Train". *Newsrail* (Australian Railway Historical Society (Victorian Division)): pages 71–76. February 1998.

[2] Chris Banger (March 1997). "Rail Passenger Service Withdrawals Since 1960". *Newsrail* (Australian Railway Historical Society (Victorian Division)): pages 77–82.

[3] Fiddian, Mark (1997). *Trains, Tracks, Travellers. A history of the Victorian Railways.* South Eastern Independent Newspapers. p. 154. ISBN 1 875475 12 5.

[4] The Victorian Transport Plan

[5] "Gippsland Public Transport Links Boosted in Budget" (http://www.doi.vic.gov.au/DOI/DOIElect.nsf/$UNIDS+for+Web+Display/493DA3882B071AD3CA25743D000CD078/$FILE/MediaRelease-SouthGippsland-020508.pdf). *Media release from the Minister for Public Transport.* www.doi.vic.gov.au. 2 May 2008. . Retrieved 2008-05-04. Communication from Minister Lynne Kosky, 03 June 2008.

[6] TransNet Logistics (May 1, 2007). "Leongatha Rail Line: Reintroduction of Passenger Train Services (Review Summary & Transport Options Assessment Report)" (http://www.doi.vic.gov.au/DOI/DOIElect.nsf/$UNIDS+for+Web+Display/754A14B7BD86361FCA25743D0005919B/$FILE/Leongatha_Rail_Line_Review.pdf). www.doi.vic.gov.au. . Retrieved 2008-05-04.

External links

- Statistics and detailed schematic map (http://www.vicsig.net/index.php?page=infrastructure&section=lineguide&line=Leongatha) at the vicsig (http://www.vicsig.net/) enthusiast website
- Photos: Between Cranbourne and Clyde stations (http://photosbymerlin.fotopic.net/c921878.html)

Pakenham_railway_line

Pakenham railway line, Melbourne	
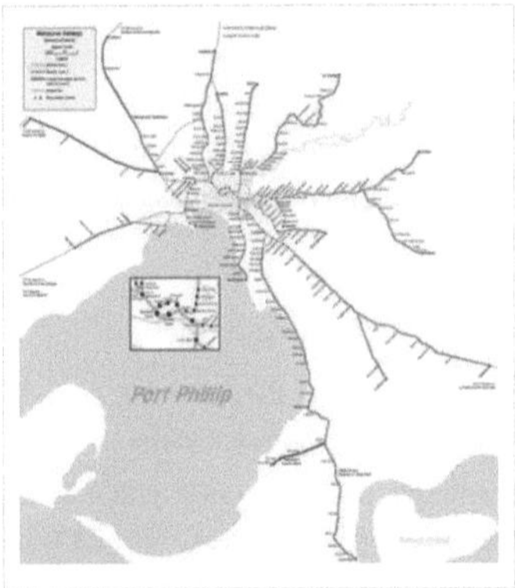	
Line details	
Commenced	1877
Completed	1879
Length	56.9 km (35.4 mi)
Stations	29 (via Loop)
Tracks	Quadruplicated to Caufield, remainder double track
Used by	Metro Trains Melbourne, V/Line and freight to Gippsland
Service pattern	Stopping all stations, some
Rolling stock	Hitachi, Comeng, Siemens
Connections	Sandringham, Frankston, and Cranbourne lines
Former connections	Rosstown Railway, Outer Circle, and Spring Vale Cemetery lines
Railways in Melbourne	
List of stations **Freight railways** **Closed railways** **Proposed railways**	

The **Pakenham railway line** in Melbourne, Australia is the suburban electrified portion of the railway to Bairnsdale in eastern Gippsland. In addition to suburban electric trains, it carries diesel passenger trains and freight trains to Gippsland.

Description

The Pakenham line traverses mainly flat country with few curves and fairly minimal earthworks for most of the line. However between South Yarra and Malvern the line has been lowered into a cutting to eliminate level crossings, and between Malvern and Caulfield it has been raised on an embankment for the same reason. Also, it encounters some gentle hills between Narre Warren and Beaconsfield which require some more extensive earthworks to negotiate. After Caulfield, the line has numerous level crossings, as well as some bridges over and under roads.

Most of the line goes through built-up suburbs and some industrial areas, but after Dandenong, it gets into more open countryside, passing by open fields and farms, particularly after Beaconsfield. This outer portion of the line is one of Melbourne's main growth corridors, which is rapidly replacing the farms with houses and adding many new riders to the line each year.

Infrastructure

The Pakenham line branches from the Lilydale line at Richmond and shares four tracks with the Frankston line to Caulfield, running next to the Sandringham line as far as South Yarra.

From Caulfield the rest of the line is double track, with three platforms at Oakleigh, Dandenong and Westall. Although there are plans to add a third track on some sections between Caulfield and Dandenong as part of the Dandenong railway line triplication project.[1] [2] [3] Automatic block signalling is provided throughout.

Intermediate terminating facilities are provided at Caulfield, Oakleigh, Westall, Springvale, Dandenong, Narre Warren and Berwick. Stabling sidings for suburban trains are provided at Caulfield, Oakleigh, Westall, Dandenong, and Pakenham, although those at Caulfield and Oakleigh are not normally used. Westall is also the location of a suburban train maintenance workshop.

History

The Pakenham line began as part of the main line to Gippsland, but at the time the route through the suburbs has not been determined, so the first portion to open was the section between Oakleigh and Bunyip (Gippsland) in October 1877. Oakleigh was connected to the city at South Yarra in April 1879.

The section from South Yarra to Hawksburn was duplicated from opening, and duplication was extended to Caulfield in 1881, Oakleigh in 1883, and Dandenong in 1891.

In 1915 the line from South Yarra to Caulfield was quadrupled, and it was possibly at the same time that this section of the line was lowered into a cutting to eliminate numerous level crossings. Power signalling was provided between Richmond and Hawksburn at the same time, then on to Caulfield in 1921.

Electrification of the line to Dandenong occurred in two stages in 1922. Power signalling was extended to Carnegie in 1933 and Oakleigh in 1940.

In 1954 the line beyond Dandenong was electrified mainly because of the expected briquette traffic from the brown coal mines in the Latrobe Valley, and over the next two years most of the line between Dandenong and Pakenham was duplicated and provided with power signalling, although Narre Warren to Berwick was not done until 1962.

Oakleigh to Dandenong was converted to power signalling in stages from 1970 to 1972.

In 1975, suburban services were extended beyond Dandenong to Pakenham, which until then had only been served by the passenger trains connecting Gippsland to Melbourne.

In 2002, General Motors station closed and all trains ran express between Hallam and Dandenong.

On 1 March 2007, with the abolition of Zone 3, Pakenham was re-zoned to Zone 2 in the Melbourne Metropolitan rail system. This brought the cost of train fares down, improving accessibility to the public.

Services

- Pakenham, Officer, Beaconsfield, Berwick, Narre Warren, Hallam, Dandenong, all stations to Caulfield/Malvern, express to South Yarra, then all stations to Flinders Street via City Loop. During peak hour, these services commence from Berwick, Dandenong, Westall or Oakleigh.
- Pakenham, Officer, Beaconsfield, Berwick, Narre Warren, Hallam, Dandenong, Noble Park, Sandown Park, Springvale, Clayton, Oakleigh, Caulfield, South Yarra, then all stations to Flinders Street via City Loop. Peak only.

Line guide

Bold stations are termini, where some train services terminate; *italic* stations are staffed; and stations with an asterisk (*) are manned only during morning peak.

Branches from the City Loop at ***Flinders Street*** and *Parliament.*

References

[1] "Plan for new Dandenong train line off the rails" by Jason Dowling, The Age, August 5, 2008. (http://www.theage.com.au/national/plan-for-new-dandenong-train-line-off-the-rails-20080804-3pxv.html)

[2] "Anger at Caulfield end of line", by Paul Riordan, Caulfield Glen Eira Leader, August 5, 2008 (http://www.caulfieldgleneiraleader.com.au/article/2008/08/05/40455_ckv_news.html)

[3] Dandenong Rail Corridor Project, Department of Transport website (http://www.doi.vic.gov.au/DOI/Internet/transport.nsf/AllDocs/7D244319338C6A0ACA2572FA0005B349?OpenDocument)

External links

- Timetables (http://www.metlinkmelbourne.com.au/route/view/11)
- Official line map (http://www.metlinkmelbourne.com.au/maps_stations_stops/metropolitan_trains/pakenham_cranbourne_lines)
- Network map (http://www.metlinkmelbourne.com.au/maps_stations_stops/metropolitan_trains)
- Statistics and detailed schematic map (http://www.vicsig.net/index.php?page=infrastructure&section=lineguide&line=Pakenham) at the VicSig (http://www.vicsig.net/) enthusiast website

Rail_transport

BNSF Railway freight service in the United States

Rail transport
Operations
Track
Maintenance
High-speed
Gauge
Stations
Trains
Locomotives
Rolling stock
Railways
History
Attractions
Terminology
By country
Accidents
Modelling

Rail transport is a means of conveyance of passengers and goods by way of wheeled vehicles running on rail tracks. In contrast to road transport, where vehicles merely run on a prepared surface, rail vehicles are also directionally guided by the tracks they run on. Track usually consists of steel rails installed on sleepers/ties and ballast, on which the rolling stock, usually fitted with metal wheels, moves. However, other variations are also possible, such as slab track where the rails are fastened to a concrete foundation resting on a prepared subsurface.

Rolling stock in railway transport systems generally has lower frictional resistance when compared with highway vehicles, and the passenger and freight cars (carriages and wagons) can be coupled into longer trains. The operation is carried out by a railway company, providing transport between train stations or freight customer facilities. Power is provided by locomotives which either draw electrical power from a railway electrification system or produce their own power, usually by diesel engines. Most tracks are accompanied by a signalling system. Railways are a safe land transport system when compared to other forms of transport.[1] Railway transport is capable of high levels of passenger and cargo utilization and energy efficiency, but is often less flexible and more capital-intensive than highway transport is, when lower traffic levels are considered.

The oldest, man-hauled railways date to the 6th century B.C, with Periander, one of the Seven Sages of Greece, credited with its invention. With the British development of the steam engine, it was possible to construct mainline railways, which were a key component of the industrial revolution. Also, railways reduced the costs of shipping, and allowed for fewer lost goods. The change from canals to railways allowed for "national markets" in which prices varied very little from city to city. Studies have shown that the invention and development of the railway in Europe was one of the most important technological inventions of the late 19th century for the United States, without which, GDP would have been lower by 7.0% in 1890. In the 1880s, electrified trains were introduced, and also the first tramways and rapid transit systems came into being. Starting during the 1940s, the non-electrified railways in most countries had their steam locomotives replaced by diesel-electric locomotives, with the process being almost complete by 2000. During the 1960s, electrified high-speed railway systems were introduced in Japan and a few other countries. Other forms of guided ground transport outside the traditional railway definitions, such as monorail or maglev, have been tried but have seen limited use.

History

The history of the growth, decline and resurgence of rail transport can be divided up into several discrete periods defined by the principal means of motive power used.

Pre-steam

The earliest evidence of a railway was a 6-kilometre (3.7 mi) Diolkos wagonway, which transported boats across the Corinth isthmus in Greece during the 6th century BC. Trucks pushed by slaves ran in grooves in limestone, which provided the track element. The Diolkos ran for over 600 years.[2]

Horsecar in Brno, Czech Republic

Railways began reappearing in Europe after the Dark Ages. The earliest known record of a railway in Europe from this period is a stained-glass window in the Minster of Freiburg im Breisgau in Germany, dating from around 1350.[3] In 1515, Cardinal Matthäus Lang wrote a description of the Reisszug, a funicular railway at the Hohensalzburg Castle in Austria. The line originally used wooden rails and a hemp haulage rope, and was operated by human or animal power. The line still exists, albeit in updated form, and is one of the oldest railways still to operate.[4] [5]

By 1550, narrow gauge railways with wooden rails were common in mines in Europe.[6] By the 17th century, wooden wagonways were common in the United Kingdom for transporting coal from mines to canal wharfs for transshipment to boats. The world's oldest working railway, built in 1758, is the Middleton Railway in Leeds. In 1764, the first gravity railroad in the United States was built in Lewiston, New York.[7] The first permanent tramway was the Leiper Railroad in 1810.[8]

The first iron plate rail way made with cast iron plates on top of wooden rails, was taken into use in 1768.[9] This allowed a variation of gauge to be used. At first only balloon loops could be used for turning, but later, movable points were taken into use that allowed for switching.[10] From the 1790s, iron edge rails began to appear in the United Kingdom.[11] In 1803, William Jessop opened the Surrey Iron Railway in south London, arguably the world's first horse-drawn public railway.[12] The invention of the wrought iron rail by John Birkinshaw in 1820 allowed the short, brittle, and often uneven, cast iron rails to be extended to 15 feet (4.6 m) lengths.[13] These were succeeded by steel in 1857.[11]

The railroad era in the United States began in 1830 when Peter Cooper's locomotive, *Tom Thumb*, first steamed along 13 miles (21 km) of Baltimore and Ohio railroad track.[14] In 1833 the nation's second railroad ran 136 miles (219 km) from Charleston to Hamburg in South Carolina.[15] Not until the 1850s, though, did railroads offer long distance service at reasonable rates. A journey from Philadelphia to Charleston involved eight different gauges, which meant that passengers and freight had to change trains seven times. Only at places like Bowling Green, Kentucky, the railroads were connected to one another.

Age of steam

A British steam locomotive-hauled train

The development of the steam engine during the Industrial revolution in the United Kingdom spurred ideas for mobile steam locomotives that could haul trains on tracks. James Watt's patented steam engines of 1769 (revised in 1782) were heavy low-pressure engines which were not suitable for use in locomotives. However, in 1804, using high-pressure steam, Richard Trevithick demonstrated the first locomotive-hauled train in Merthyr Tydfil, United Kingdom.[16] [17] Accompanied with Andrew Vivian, it ran with mixed success,[18] breaking some of the brittle cast-iron plates.[19] Two years later, the first passenger horse-drawn railway was opened nearby between Swansea and Mumbles.[20]

The earliest British steam railways

In 1811, John Blenkinsop designed the first successful and practical railway locomotive[21] —a rack railway worked by a steam locomotive between Middleton Colliery and Leeds on the Middleton Railway. The locomotive, *Salamanca*, was built the following year.[22] :20 In 1825, George Stephenson built the *Locomotion* for the Stockton and Darlington Railway, north east England, which was the first public steam railway in the world. In 1829, he built *The Rocket* which was entered in and won the Rainhill Trials. This success led to Stephenson establishing his company as the pre-eminent builder of steam locomotives used on railways in the United Kingdom, the United States and much of Europe.[22] :24–30

In 1830, the first intercity railway, the Liverpool and Manchester Railway, opened. The gauge was that used for the early wagonways and had been adopted for the Stockton and Darlington Railway.[23] The 1435 mm (4 ft 8 $^1/_2$ in) width became known as the international standard gauge, used by about 60% of the world's railways. This spurred the spread of rail transport outside the UK.

By the early 1850s Britain had over 7,000 miles of railway, 'a stunning achievment given that only twenty years had elapsed since the opening of the Liverpool and Manchester Railway.[19]

Early railroads in the USA

Railroads (as they were known in the USA) were built on a far larger scale than those in Continental Europe, both in terms of the distances covered and also in the loading gauge adopted which allowed for heavier locomotives and double-deck trains.

The Baltimore and Ohio that opened in 1830 was the first to evolve from a single line to a network in the United States.[24] By 1831, a steam railway connected Albany and Schenectady, New York, a distance of 16 miles, which was covered in 40 minutes.[25]

The years between 1850 and 1890 saw phenomenal growth in the US railroad system, which at its peak constituted one third of the world's total mileage.[26] Although the American Civil War placed a temporary halt to major new developments the conflict did demonstrate the enormous strategic importance of railways at times of war. After the war major developments include the first elevated railway built in New York in 1867 and the symbolically important first transcontinental railway was completed in 1869.[27]

Electrification and dieselisation

0-Series Shinkansen, introduced in 1964, triggered the intercity train travel boom.

Elevated section of the Chicago 'L'

Experiments with electrical railways were started by Robert Davidson in 1838. He completed a battery-powered carriage capable of 6.4 km/h (4 mph). The Gross-Lichterfelde Tramway was the first to use electricity fed to the trains en-route, when it opened in 1881. Overhead wires were taken into use in the Mödling and Hinterbrühl Tram in Austria in October 1883. At first, this was taken into use on tramways that, until then, had been horse-drawn tramcars. The first conventional completely electrified railway mainline was the 106 km Valtellina line in Italy that was opened on 4 September 1902. During the 1890s, many large cities, such as London, Paris and New York used the new technology to build rapid transit for urban commuting. In smaller cities, tramways became common and were often the only mode of public transport until the introduction of buses in the 1920s. In North America, interurbans became a common mode to reach suburban areas. At first, all electric railways used direct current but, in 1904, the Stubaital Line in Austria opened with alternating current.[28]

Steam locomotives require large pools of labour to clean, load, maintain and run. After World War II, dramatically increased labour costs in developed countries made steam an increasingly costly form of motive power. At the same time, the war had forced improvements in internal combustion engine technology that made diesel locomotives cheaper and more powerful. This caused many railway companies to initiate programmes to convert all unelectrified sections from steam to diesel locomotion.

Following the large-scale construction of motorways after the war, rail transport became less popular for commuting and air transport started taking large market shares from long-haul passenger trains. Most tramways were either replaced by rapid transits or buses, while high transshipment costs caused short-haul freight trains to become uncompetitive. The 1973 oil crisis led to a change of mind set and most tram systems that had survived into the 1970s remain today. At the same time, containerization allowed freight trains to become more competitive and participate in intermodal freight transport. With the 1964 introduction of the Shinkansen high-speed rail in Japan, trains could again have a dominant position on intercity travel. During the 1970s, the introduction of automated rapid transit systems allowed cheaper operation. The 1990s saw an increased focus on accessibility and low-floor trains. Many tramways have been upgraded to light rail and many cities that closed their old tramways have reopened new light railway systems.

Luas in Dublin, Ireland

Innovations

Many benchmarks in equipment and infrastructure led to the growing use of railways. Some innovative features taking place in the 19th and 20th centuries included wood cars replaced with all-steel cars, which provided better safety and maintenance; iron rails replaced with steel rails, which provided higher speed and capacity with lower weight and cost; stove-heated cars to steam-heating cars, piped from locomotive; gas lighting to electric lighting, with use of battery/alternator unit beneath the car; development of air-conditioning with additional underbody equipment and ice compartment. Some innovative rolling stock included the lightweight, diesel-powered streamliner, which was an modernistic, aerodynamically-styled train with flowing contours; then came the ultra-lightweight car with internal combustion engine in each train's power car; others included the dome car, turbined-powered trains, bilevel rolling stock, and the high-tech/high-speed electric trains.[29]

Even more, in the first half of the 20th century, infrastructure elements adopted technological changes including the continuously welded rail that was 1/4 miles long; concrete tie usage; double tracking major lines; intermodal terminal and handling technology; advances in diesel-electric propulsion to include AC traction systems and propulsion braking systems; and just-in-time inventory control. Beyond technology, even management of systems seen improvements with the adoption of environmental impact concerns; heightened concern of employee and public safety; introduction of urban area rail networks and public agencies to manage them; and downsizing of the industry employment with greater use of contractors and consultants.[30]

Trains

A train is a connected series of rail vehicles that move along the track. Propulsion for the train is provided by a separate locomotive or from individual motors in self-propelled multiple units. Most trains carry a revenue load, although non-revenue cars exist for the railway's own use, such as for maintenance-of-way purposes. The engine driver controls the locomotive or other power cars, although people movers and some rapid transits are driverless.

Haulage

Russian 2TE10U diesel locomotive

Traditionally, trains are pulled using a locomotive. This involved a single or multiple powered vehicles being located at the front of the train and providing sufficient adhesion to haul the weight of the full train. This remains dominant for freight trains and is often used for passenger trains. A push-pull train has the end passenger car equipped with a driver's cab so the engine driver can remotely control the locomotive. This allows one of the locomotive-hauled trains drawbacks to be removed, since the locomotive need not be moved to the end of the train each time the train changes direction. A railroad car is a vehicle used for the haulage of either passengers or freight.

A multiple unit has powered wheels throughout the whole train. These are used for rapid transit and tram systems, as well as many both short- and long-haul passenger trains. A railcar is a single, self-powered car. Multiple units have a driver's cab at each end of the unit and were developed following the ability to build electric motors and engines small enough to build under the coach. There are only a few freight multiple units, most of which are high-speed post trains.

Motive power

A RegioSwinger multiple unit of the Croatian Railways

Steam locomotives are locomotives with a steam engine that provides adhesion. Coal, petroleum, or wood is burned in a firebox. The heat boils water in the fire-tube boiler to create pressurized steam. The steam travels through the smokebox before leaving via the chimney. In the process, it powers a piston that transmits power directly through a connecting rod (US: main rod) and a crankpin (US: wristpin) on the driving wheel (US main driver) or to a crank on a driving axle. Steam locomotives have been phased out in most parts of the world for economical and safety reasons although many are preserved in working order by heritage railways.

Electric locomotives draw power from a stationary source via an overhead wire or third rail. Some also or instead use a battery. A transformer in the locomotive converts the high voltage, low current power to low voltage, high current used in the electric motors that power the wheels. Modern locomotives use three-phase AC induction motors. Electric locomotives are the most powerful traction.

They are also the cheapest to run and provide less noise and no local air pollution. However, they require high capital investments both for the overhead lines and the supporting infrastructure. Accordingly, electric traction is used on urban systems, lines with high traffic and for high-speed rail.

Diesel locomotives use a diesel engine as the prime mover. The energy transmission may be either diesel-electric, diesel-mechanical or diesel-hydraulic but diesel-electric is dominant. Electro-diesel locomotives are built to run as diesel-electric on unelectrified sections and as electric locomotives on electrified sections.

Alternative methods of motive power include magnetic levitation, horse-drawn, cable, gravity, pneumatics and gas turbine.

Passenger trains

Interior view of the top deck of a VR InterCity2 double-deck carriage

A passenger train travels between stations where passengers may embark and disembark. The oversight of the train is the duty of a guard/train manager. Passenger trains are part of public transport and often make up the stem of the service, with buses feeding to stations. Passenger trains can involve a variety of functions including long distance intercity travel, daily commuter trips, or local urban transit services. They even include a diversity of vehicles, operating speeds, right of way requirements, and service frequency. Passenger trains usually can be divided into two operations: intercity railway and intracity transit. Whereas as intercity railway involve higher speeds, longer routes, and lower frequency (usually scheduled), intracity transit involves lower speeds, shorter routes, and higher frequency (especially during peak hours).[30]

Intercity trains are long-haul trains that operate with few stops between cities. Trains typically have amenities such as a dining car. Some lines also provide over-night services with sleeping cars. Some long-haul trains been given a specific name. Regional trains are medium distance trains that connect cities with outlying, surrounding areas, or provide a regional service, making more stops and having lower speeds. Commuter trains serve suburbs of urban areas, providing a daily commuting service. Airport rail links provide quick access from city centres to airports.

High-speed rail are special inter-city trains that operates at much higher speeds than conventional railways, the limit being regarded at 200 to 320 km/h. High-speed trains are used mostly for long-haul service and most systems are in Western Europe and East Asia. The speed record is 574.8 km/h (357.2 mph), set by a modified French TGV.[31] [32] Magnetic levitation trains such as the Shanghai airport train use under-riding magnets which attract themselves upward towards the underside of a guideway and this line has achieved somewhat higher peak speeds in day-to-day operation than conventional high-speed railways, although only over short distances. Due to their heightened speeds, route alignments for high-speed rail tend to be steeper grades and broader curves compared to conventional railways. Their high kinetic energy translates to higher horsepower-to-ton ratios (20 hp/ton); this allows trains to accelerate and maintain higher speeds and negotiate steep grades as momentum builds up and recovered in downgrades (reducing cut, fill, and tunneling requirements). Since lateral forces act on curves, curvatures are designed with the highest possible radius. All these features are dramatically different from freight operations, thus justifying exclusive high-speed rail lines if it is economically feasible.[30]

Rapid transit is an intracity system built in large cities and has the highest capacity of any passenger transport system. It is grade separated and commonly built underground or elevated. At street level, smaller trams can be used. Light rails are upgraded trams that have step-free access, their own right-of-way and sometimes sections underground. Monorail systems operate as elevated, medium capacity systems. A people mover is a driverless, grade-separated train that serves only a few stations, as a shuttle. Due to the wide variety of rapid transit systems without much uniformity, route alignment vary widely with diverse right-of-ways (private land, side of road, street median) and geometric characteristics (sharp or broad curves, steep or gentle grades). For instance, the Chicago El trains are designed with extremely short cars to negotiate the sharp curves in the Loop. NJ's PATH have similar-sized cars to accommodate curves in the trans-Hudson tunnels. San Francisco's BART operate large cars on its well-engineered routes.[30]

Freight train

Bulk cargo of minerals

A freight train hauls cargo using freight cars specialized for the type of goods. Freight trains are very efficient, with economy of scale and high energy efficiency. However, their use can be reduced by lack of flexibility, if there is need of transshipment at both ends of the trip due to lack of tracks to the points of pick-up and delivery. Authorities often encourage the use of cargo rail transport due to its environmental profile.

Container trains have become the dominant type in the US for non-bulk haulage. Containers can easily be transshipped to other modes, such as ships and trucks, using cranes. This has succeeded the boxcar (wagon-load), where the cargo had to be loaded and unloaded into the train manually. The intermodal containerization of cargo has revolutionized the supply chain logistics industry, reducing ship costs significantly. In Europe, the sliding wall wagon has largely superseded the ordinary covered wagons. Other types of cars include refrigerator cars, stock cars for livestock and autoracks for road vehicles. When rail is combined with road transport, a roadrailer will allow trailers to be driven onto the train, allowing for easy transition between road and rail.

Bulk handling represents a key advantage for rail transport. Low or even zero transshipment costs combined with energy efficiency and low inventory costs allow trains to handle bulk much cheaper than by road. Typical bulk cargo includes coal, ore, grains and liquids. Bulk is transported in open-topped cars, hopper cars and tank cars.

Infrastructure

Right of way

Railway tracks are laid upon land owned or leased by the railway company. Owing to the desirability of maintaining modest grades, rails will often be laid in circuitous routes in hilly or mountainous terrain. Route length and grade requirements can be reduced by the use of alternating cuttings, bridges and tunnels—all of which can greatly increase the capital expenditures required to develop a right of way, while significantly reducing operating costs and allowing higher speeds on longer radius curves. In densely urbanized areas, railways are sometimes laid in tunnels to minimize the effects on existing properties.

Trackage

Track consists of two parallel steel rails, anchored perpendicular to members called ties (sleepers) of timber, concrete, steel, or plastic to maintain a consistent distance apart, or rail gauge. Rail gauges are usually categorised as Standard gauge (1435 mm (4 ft 8 $^1/_2$ in)) used on approximately 60% of the world's existing railway lines, Broad gauge and Narrow gauge. In addition to the rail gauge, the tracks will be laid to conform with a Loading gauge which defines the maximum height and width for railway vehicles and their loads to ensure safe passage through bridges, tunnels and other structures.

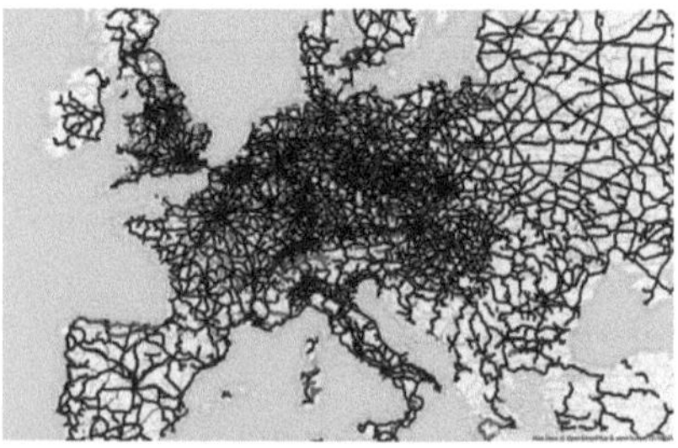

Map of railways in Europe with main operational lines shown in black, heritage railway lines in green and former routes in light blue

The track guides the conical, flanged wheels, keeping the cars on the track without active steering and therefore allowing trains to be much longer than road vehicles. The rails and ties are usually placed on a foundation made of compressed earth on top of which is placed a bed of ballast to distribute the load from the ties and to prevent the track from buckling as the ground settles over time under the weight of the vehicles passing above.

The ballast also serves as a means of drainage. Some more modern track in special areas is attached by direct fixation without ballast. Track may be prefabricated or assembled in place. By welding rails together to form lengths of continuous welded rail, additional wear and tear on rolling stock caused by the small surface gap at the joints between rails can be counteracted; this also makes for a quieter ride (passenger trains).

Long freight train crossing the Stoney Creek viaduct on the Canadian Pacific Railway in southern British Columbia

On curves the outer rail may be at a higher level than the inner rail. This is called superelevation or cant. This reduces the forces tending to displace the track and makes for a more comfortable ride for standing livestock and standing or seated passengers. A given amount of superelevation will be the most effective over a limited range of speeds.

Turnouts, also known as points and switches, are the means of directing a train onto a diverging section of track. Laid similar to normal track, a point typically consists of a frog (common crossing), check rails and two switch rails. The switch rails may be moved left or right, under the control of the signalling system, to determine which path the train will follow.

Spikes in wooden ties can loosen over time, but split and rotten ties may be individually replaced with new wooden ties or concrete substitutes. Concrete ties can also develop cracks or splits, and can also be replaced individually. Should the rails settle due to soil subsidence, they can be lifted by specialized machinery and additional ballast tamped under the ties to level the rails.

Periodically, ballast must be removed and replaced with clean ballast to ensure adequate drainage. Culverts and other passages for water must be kept clear lest water is impounded by the trackbed, causing landslips. Where trackbeds are placed along rivers, additional protection is usually placed to prevent streambank erosion during times of high water. Bridges require inspection and maintenance, since they are subject to large surges of stress in a short period of time when a heavy train crosses.

Train Inspection Systems

The inspection of railway equipment is essential for the safe movement of trains. Many types of defect detectors are in use on the world's railroads. These devices utilize technologies vary from a simplistic paddle and switch to infrared and laser scanning, and even ultrasonic audio analysis. Their use has avoided many rail accidents over the 70 years they have been used.

A Hot bearing detector w/ dragging equipment unit

Signalling

Railway signalling is a system used to control railway traffic safely to prevent trains from colliding. Being guided by fixed rails with low friction, trains are uniquely susceptible to collision since they frequently operate at speeds that do not enable them to stop quickly or within the driver's sighting distance. Most forms of train control involve movement authority being passed from those responsible for each section of a rail network to the train crew. Not all methods require the use of signals, and some systems are specific to single track railways.

The signalling process is traditionally carried out in a signal box, a small building that houses the lever frame required for the signalman to operate switches and signal equipment. These are placed at various intervals along the route of a railway, controlling specified sections of track. More recent technological developments have made such operational doctrine superfluous, with the centralization of signalling operations to regional control rooms. This has been facilitated by the increased use of computers, allowing vast sections of track to be monitored from a single location. The common method of block signalling divides the track into zones guarded by combinations of block signals, operating rules, and automatic-control devices so that only one train may be in a block at any time.

Electrification

The electrification system provides electrical energy to the trains, so they can operate without a prime mover onboard. This allows lower operating costs, but requires large capital investments along the lines. Mainline and tram systems normally have overhead wires, which hang from poles along the line. Grade-separated rapid transit sometimes use a ground third rail.

Power may be fed as direct or alternating current. The most common DC voltages are 600 and 750 V for tram and rapid transit systems, and 1,500 and 3,000 V for mainlines. The two dominant AC systems are 15 kV AC and 25 kV AC.

Stations

A railway station serves as an area where passengers can board and alight from trains. A goods station is a yard which is exclusively used for loading and unloading cargo. Large passenger stations have at least one building providing conveniences for passengers, such as purchasing tickets and food. Smaller stations typically only consist of a platform. Early stations were sometimes built with both passenger and goods facilities.[33]

Platforms are used to allow easy access to the trains, and are connected to each other via underpasses, footbridge and level crossings. Some large stations are built as cul-de-sac, with trains only operating out from one direction. Smaller stations normally serve local residential areas, and may have connection to feeder bus services. Large stations, in particular central stations, serve as the main public transport hub for the city, and have transfer available between rail services, and to rapid transit, tram or bus services.

Records

- The total number of Indian Railways stations is thought to be near about 8000.

Operations

Ownership

Traditionally, the infrastructure and rolling stock are owned and operated by the same company. This has often been by a national railway, while other companies have had private railways. Since the 1980s, there has been an increasing tendency to split up railway companies, with separate companies owning the stock from those owning the infrastructure, particularly in Europe, where this is required by the European Union. This has allowed open access by any train operator to any portion of the European railway network.

In the United States, railways, such as Union Pacific, are privately owned

In the U.S., virtually all rail networks and infrastructure are privately-owned with passenger lines, primarily Amtrak, operating as tenants on the freight lines. Consequently, operations must be closely synchronized and coordinated between freight and passenger railways for timeliness of their schedules. Due to this shared system, both are regulated by the Federal Railroad Administration (FRA) and follow the AREMA standards for track work and AAR standards for vehicles.[30]

Financing

The main source of income for railway companies is from ticket revenue (for passenger transport) and shipment fees for cargo. Discounts and monthly passes are sometimes available for frequent travellers. Freight revenue may be sold per container slot or for a whole train. Sometimes, the shipper owns the cars and only rents the haulage. For passenger transport, advertisement income can be significant.

Government may choose to give subsidies to rail operation, since rail transport has fewer externalities than other dominant modes of transport. If the railway company is state-owned, the state may simply provide direct subsidies in exchange for an increased production. If operations have been privatized, several options are available. Some countries have a system where the infrastructure is owned by a government agency or company—with open access to the tracks for any company that meets safety requirements. In such cases, the state may choose to provide the tracks free of charge, or for a fee that does not cover all costs. This is seen as analogous to the government providing free access to roads. For passenger operations, a direct subsidy may be paid to a public-owned operator, or public service obligation tender may be helt, and a time-limited contract awarded to the lowest bidder.

U.S.'s national passenger rail service, Amtrak, is a private railroad company chartered by the government. Similarly, Canada's VIA Rail system operates in the same fashion. As private passenger services lost significant ground in competition to the automobile and airplane and was forced out the market, they became stockholders of Amtrak either with a cash entrance fee or relinquishing their locomotives and rolling stock. Government aid supports Amtrak by supplying start-up capital and makes up for loses at end of the fiscal year.[29]

Safety

Rail transport is one of the safest forms of land travel.[34] Trains can travel at very high speed, but they are heavy, are unable to deviate from the track and require a great distance to stop. Possible accidents include derailment (jumping the track), a collision with another train or collision with an automobile or other vehicle at level crossings. The latter accounts for the majority of rail accidents and casualties. The most important safety measures to prevent accidents are strict operating rules, e.g. railway signalling and gates or grade separation at crossings. Train whistles, bells or horns warn of the presence of a train, while trackside signals maintain the distances between trains.

Train crash at Montparnasse Station, Paris, France, in 1895

An important element in the safety of many high-speed inter-city networks such as Japan's Shinkansen is the fact that trains only run on dedicated railway lines, without level crossings. This effectively eliminates the potential for collision with automobiles, other vehicles and pedestrians, vastly reduces the likelihood of collision with other trains and helps ensure services remain timely.

Maintenance

As in any infrastructure asset, railways must keep up with periodic inspection and maintenance in order to minimize effect of infrastructure failures that can disrupt freight revenue operations and passenger services. Because passengers are considered the most *crucial cargo* and usually operate at higher speeds, steeper grades, and higher capacity/frequency, their lines are especially important. Inspection practices include track geometry cars or walking inspection. Curve maintenance especially for transit services includes gauging, fastener tightening, and rail replacement. Rail corrugation is a common issue with transit systems due to the high number of light-axle, wheel passages which result in grinding of the wheel/rail interface. Since maintenance may overlap with operations, maintenance windows (nighttime hours, off-peak hours, altering train schedules or routes) must be closely followed. In addition, passenger safety during maintenance work (inter-track fencing, proper storage of materials, track work notices, hazards of equipment near states) must be regarded at all times. At times, maintenance access problems can emerge due to tunnels, elevated structures, and congested cityscapes. Here, specialized equipment, smaller versions of conventional maintenance gear are used.[30]

Unlike highways or road networks where capacity is disaggregated into unlinked trips over individual route segments, railway capacity is fundamentally considered a network system. As a result, many components are causes and effects of system disruptions. Maintenance must acknowledge the vast array of a route's performance (type of train service, origination/destination, seasonal impacts), line's capacity (length, terrain, number of tracks, types of train control), train's throughput (max speeds, acceleration/deceleration rates), and service features with shared passenger-freight tracks (sidings, terminal capacities, switching routes, and design type).[30]

Impact

Energy

Rail transport is an energy-efficient [35] but capital-intensive, means of mechanized land transport. The tracks provide smooth and hard surfaces on which the wheels of the train can roll with a minimum of friction. Moving a vehicle on and/or through a medium (land, sea, or air) requires overcoming resistance to motion. A land vehicle's total resistance (in pounds or Newtons) is a quadratic function of the vehicle's speed:

BNSF Railway freight service in the United States

where:

> R denotes total resistance
>
> a denotes initial constant resistance
>
> v denotes speed
>
> c denotes shape, frontal area, and sides of vehicle
>
> v^2 denotes effect of air resistance[30]

Essentially, resistance differs between vehicle's contact point and surface of roadway. Metal wheels on metal rails have a significant advantage of overcoming resistance compared to rubber-tired wheels on any road surface (railway - 0.001g at 10 mph and 0.024g at 60 mph; truck - 0.009g at 10 mph and 0.090 at 60 mph). In terms of cargo capacity combining speed and size being moved in a day (ton-miles/day):

German InterCityExpress

- human - can carry 100 lbs. 20 miles per day (one ton-mile/day)
- horse and wheelbarrow - can carry 4 ton-miles/day
- horse cart on good pavement - can carry 10 ton-miles/day
- fully utility truck - can carry 20,000 ton-miles/day
- long-haul train - can carry 500,000 ton-miles/day[30]

In terms of motive power, the horsepower and weight ratio, used to overcome resistance to motion when locomotives convert fuel to heat for propulsion a slow-moving barge requires 0.2 hp/net ton, a railway and pipeline requires 2.5 hp/net ton, and truck requires 10 hp/net ton. However, at higher speeds, a railway overcomes the barge and proves most economical.[30]

As an example, a typical modern wagon can hold up to 113 tonnes of freight on two four-wheel bogies. The contact area between each wheel and the rail is a strip no more than a few millimetres wide, which minimizes friction. The track distributes the weight of the train evenly, allowing significantly greater loads per axle and wheel than in road transport, leading to less wear and tear on the permanent way. This can save energy compared with other forms of transport, such as road transport, which depends on the friction between rubber tires and the road. Trains have a small frontal area in relation to the load they are carrying, which reduces air resistance and thus energy usage.

In addition, the presence of track guiding the wheels allows for very long trains to be pulled by one or a few engines and driven by a single operator, even around curves, which allows for economies of scale in both manpower and energy use; by contrast, in road transport, more than two articulations causes fishtailing and makes the vehicle unsafe.

Usage

Due to these benefits, rail transport is a major form of passenger and freight transport in many countries. In India, China, South Korea and Japan, many millions use trains as regular transport. It is widespread in European countries. Freight rail transport is widespread and heavily used in North America, but intercity passenger rail transport on that continent is relatively scarce outside the Northeast Corridor due to the loss of competition to other preferred modes, particular automobiles and airplanes.[29] [36]

Railway tracks running through Stanhope, United Kingdom

Africa and South America have some extensive networks such as in South Africa, Northern Africa and Argentina; but some railways on these continents are isolated lines. Australia has a generally sparse network befitting its population density, but has some areas with significant networks, especially in the southeast. In addition to the previously existing east-west transcontinental line in Australia, a line from north to south has been constructed. The highest railway in the world is the line to Lhasa, in Tibet,[37] partly running over permafrost territory. The western Europe region has the highest railway density in the world, and has many individual trains which operate through several countries despite technical and organizational differences in each national network.

Social/Economic Benefits

Railways have also been shown to contribute to social vibrancy and economic competitiveness in its ability to transport large amounts of customers and workers to city centers and inner suburbs (i.e. Washington DC as a cultural/policy center due to metrorail system, San Francisco's lively downtown due to the BART system). Hong Kong has recognized rail as "the backbone of the public transit system" and as such developed their franchised bus system and road infrastructure in compherensive alignment with their rail services.[38] China's large cities including Beijing, Shanghai, and Guangzhou recognize rail transit lines as the framework and bus lines as the main body to their metropolitan transportation systems.[39] The Japanese Shinkansen was built to meet the growing traffic demand in the "heart of Japan's industry and economy" situated on the Tokyo-Kobe line.[40]

Japanese Shinkansen

As opposed to highway expansion, indicative of the U.S. transportation policy, that incentivizes development of suburbs at the periphery, contributing to increased vehicle miles traveled, carbon emissions, development of greenfield spaces, and depletion of natural reserves, railways channel growth toward dense city agglomerations and along its artery. These sustainable arrangements revalue city spaces, local taxes, housing values, and promotion of mixed use development.[41] [42]

See also

- Environmental design in rail transportation
- International Union of Railways
- List of rail transport topics
- List of railway companies
- Megaproject
- Passenger rail terminology
- Rail transport by country
- Railway systems engineering

References

Notes

[1] According to this source (http://www.railwatch.org.uk/backtrack/rw94/rw094p06.pdf), railways are safest on both a per-mile and per-hour basis, whereas air transport is safe only on a per-mile basis

[2] Lewis, M. J. T. "Railways in the Greek and Roman World" (http://www.sciencenews.gr/docs/diolkos.pdf) (pdf). . Retrieved 11 April 2009.

[3] Hylton, Stuart (2007). *The Grand Experiment: The Birth of the Railway Age 1820-1845*. Ian Allan Publishing.

[4] Kriechbaum, Reinhard (15 May 2004). "Die große Reise auf den Berg" (http://www.die-tagespost.de/Archiv/titel_anzeige.asp?ID=8916) (in German). *der Tagespost*. . Retrieved 22 April 2009.

[5] "Der Reiszug - Part 1 - Presentation" (http://www.funimag.com/funimag10/RESZUG01.HTM). Funimag. . Retrieved 22 April 2009.

[6] Georgius Agricola (1913). *De re metallica*. ISBN 0486600068.

[7] Porter, Peter (1914). *Landmarks of the Niagara Frontier*. The Author. ISBN 0665783477.

[8] Morlok, Edward K. (11 May 2005). "First permanent railroad in the U.S. and its connection to the University of Pennsylvania" (http://www.seas.upenn.edu/~morlok/morlokpage/transp_data.html). . Retrieved 19 September 2007.

[9] at Coalbrookdale *Railways (pt 1)* (http://www.1902encyclopedia.com/R/RAI/railway-01.html). Encyclopedia Britannica. 1902. ISBN 187252463X. . Retrieved 2011-02-15.

[10] Vaughan, A. (1997). *Railwaymen, Politics and Money*. London: John Murray. ISBN 0719557461.

[11] Marshall, John (1979). *The Guiness Book of Rail Facts & Feats*. ISBN 0-900424-56-7.

[12] "Surrey Iron Railway 200th - 26th July 2003" (http://www.stephensonloco.fsbusiness.co.uk/surreyiron.htm). *Early Railways*. Stephenson Locomotive Society. . Retrieved 19 September.

[13] Skempton, A.W. (2002). *A biographical dictionary of civil engineers in Great Britain and Ireland, John Birkinshaw* (http://books.google.com/books?id=jeOMfpYMOtYC&pg=PA59). pp. 59–60. ISBN 9780727729392. .

[14] "The History of the Tom Thumb - Peter Cooper" (http://inventors.about.com/library/inventors/bl_tom_thumb.htm). Inventors.about.com. . Retrieved 8 May 2011.

[15] Storey, Steve. "Charleston & Hamburg Railroad" (http://www.railga.com/charlhmbrg.html). Railga.com. . Retrieved 8 May 2011.

[16] "Richard Trevithick's steam locomotive" (http://www.museumwales.ac.uk/en/rhagor/article/trevithic_loco/). Museumwales.ac.uk. 15 December 2008. . Retrieved 8 May 2011.

[17] "Steam train anniversary begins" (http://news.bbc.co.uk/2/hi/uk_news/wales/3509961.stm). BBC. 21 February 2004. . Retrieved 8 May 2011. "A south Wales town has begun months of celebrations to mark the 200th anniversary of the invention of the steam locomotive. Merthyr Tydfil was the location where, on 21 February 1804, Richard Trevithick took the world into the railway age when he set one of his high-pressure steam engines on a local iron master's tram rails"

[18] Payton, Philip (2004). *Oxford Dictionary of National Biography*. Oxford University Press.

[19] Chartres, J.. "Richard Trevithick". In Cannon, John. *Oxford Companion to British History*. p. 932.

[20] "Early Days of Mumbles Railway" (http://www.bbc.co.uk/wales/southwest/sites/swansea/pages/mumbles_trainanniv.shtml). BBC. 15 February 2007. . Retrieved 19 September 2007.

[21] "John Blenkinsop" (http://www.britannica.com/EBchecked/topic/69274/John-Blenkinsop). *Encyclopædia Britannica*. . Retrieved 8 May 2011.

[22] Ellis, Hamilton (1968). *The Pictorial Encyclopedia of Railways*. Hamlyn Publishing Group.

[23] "Liverpool and Manchester" (http://www.spartacus.schoolnet.co.uk/RAliverpool.htm). . Retrieved 19 September 2007.

[24] Dilts, James D. (1996). *The Great Road: The Building of the Baltimore and Ohio, the Nation's First Railroad, 1828-1853* (http://books.google.com/?id=JjrCWPwvHzIC&lpg=PR18&dq=dilts b&o&pg=PA26#v=onepage&q=first). Palo Alto, CA: Stanford University Press. p. 26. ISBN 978-0804726290. .

[25] "The Journal of Ebenezer Mattoon Chamberlain 1832-5", *Indiana Magazine of History*, Vol. XV, September, 1919, No. 3, p.233ff.

[26] Wolmar (2009) p.xiii.

[27] Ambrose, Stephen E. (2000). *Nothing Like It In The World; The men who built the Transcontinental Railroad 1863-1869* (http://books.google.com/?id=TZp_GT7PscIC&lpg=PP1&dq=ambrose nothing like it&pg=PP1#v=onepage&q=). Simon & Schuster. ISBN 0-684-84609-8. .

[28] Tokle, Bjørn (2003) (in Norwegian). *Communication gjennom 100 år*. Meldal: Chr. Salvesen & Chr. Thams's Communications Aktieselskab. p. 54.

[29] EuDaly, K, Schafer, M, Boyd, Jim, Jessup, S, McBridge, A, Glischinksi, S. (2009). The Complete Book of North American Railroading. Voyageur Press. 1-352 pgs.

[30] American Railway Engineering and Maintenance of Way Association Committee 24 - Education and Training. (2003). Practical Guide to Railway Engineering. AREMA, 2nd Ed.

[31] Associated Press (4 April 2007). "French train breaks speed record" (http://web.archive.org/web/20070407194558/http://www.cnn.com/2007/WORLD/europe/04/03/TGVspeedrecord.ap/index.html). *CNN*. Archived from the original (http://www.cnn.com/2007/WORLD/europe/04/03/TGVspeedrecord.ap/index.html) on 7 April 2007. . Retrieved 3 April 2007.

[32] Fouquet, Helene and Viscousi, Gregory (3 April 2007). "French TGV Sets Record, Reaching 357 Miles an Hour (Update2)" (http://www.bloomberg.com/apps/news?pid=newsarchive&sid=aW23Aw20nilo&refer=europe). Bloomberg L.P.. . Retrieved 8 May 2011.

[33] "The Inception of the English Railway Station". *Architectural History* (SAHGB Publications Limited) **4**: 63–76. 1961. doi:10.2307/1568245. JSTOR 1568245.

[34] U.S. Bureau of Transportation Statistics (2010). *National Transportation Statistics. Table 2-1: Transportation Fatalities by Mode* (http://www.bts.gov/publications/national_transportation_statistics/html/table_02_01.html) (Report). . Retrieved 2010-02-14.

[35] American Association of Railroads. "Railroad Fuel Efficiency Sets New Record" (http://www.progressiverailroading.com/news/article.asp?id=16740). . Retrieved 12 April 2009.

[36] "Public Transportation Ridership Statistics" (http://web.archive.org/web/20070815101950/http://www.apta.com/research/stats/ridership/). American Public Transportation Association. 2007. Archived from the original (http://www.apta.com/research/stats/ridership/) on 15 August 2007. . Retrieved 10 September 2007.

[37] "New height of world's railway born in Tibet" (http://news.xinhuanet.com/english/2005-08/24/content_3397297.htm). Xinhua News Agency. 24 August 2005. . Retrieved 8 May 2011.

[38] Hong Kong Information Services Department of the Hong Kong SAR Government. Hong Kong 2009

[39] Hau H., Yun-feng G., Zhi-gang, L., Xiao-guang, Y. (2010). Effect of Integrated Multi-Modal Transit Information on Modal Shift. Intelligent Transportation Systems (ITSC), 2010 13th International IEEE Conference. 1753-1757pg.

[40] Nishida, M., The Shinkansen High-Speed Rail Network of Japan. Proceedings of an IIASA Conference, June 27–30, 1977.

[41] Squires, G. Ed. (2002) Urban Sprawl: Causes, Consequences, & Policy Responses. The Urban Institute Press.

[42] Puentes, R. (2008). A Bridge to Somewhere: Rethinking American Transportation for the 21st Century. Brookings Institution Metropolitan Policy Report: Blueprint for American Prosperity series report.

References

pfl:Aisebahn

Victoria_(Australia)

Victoria	
 Flag	 Coat of arms
Slogan or nickname: *"Garden State"*	
Motto(s): *"Peace and Prosperity"*	
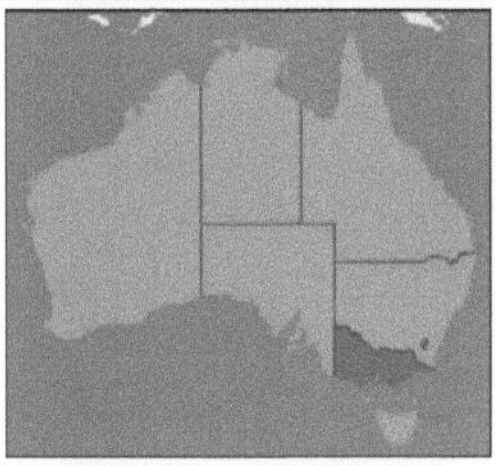 Other Australian states and territories	
Capital	Melbourne
Demonym	Victorian
Government	Constitutional monarchy
- Governor	Alex Chernov
- Premier	Ted Baillieu (LP)
Australian State	
- Established as Colony	1850
- Responsible Government	1856
- Became State	1901
- Constitution	*Constitution Act* (Vic) 1975
- *Australia Act*	3 March 1986
Area	
- Total	237629 km^2 (6th) 91749 sq mi
- Land	227416 km^2 87806 sq mi
- Water	10213 km^2 (4.3%) 3943 sq mi
Population (June 2010)	
- Population	5,547,527[1] (2nd)

- Density	24.39/km² (2nd) 63.2 /sq mi
Elevation	
- Highest	Mt Bogong 1,986 m (6,516 ft)
Gross State Product (2009–10)	
- Product ($m)	$293,313[2] (2nd)
- Product per capita	$52,872 (6th)
Time zone	UTC+10 (AEST) UTC+11 (AEDT)
Federal representation	
- House seats	37
- Senate seats	12
Abbreviations	
- Postal	VIC
- **ISO 3166-2**	AU-VIC
Emblems	
- Floral	Pink Heath[3]
- Animal	Leadbeater's possum
- Bird	Helmeted Honeyeater
- Fish	Weedy Seadragon
- Colours	Navy Blue and Silver[4]
Web site	www.vic.gov.au/ [5]

Victoria is the second most populous state in Australia.[1] Geographically the smallest mainland state, Victoria is bordered by New South Wales, South Australia, and Tasmania on Boundary Islet to the north, west and south respectively.

Victoria is Australia's most densely populated state, and has a highly centralised population, with almost 75% of Victorians living in Melbourne, the state capital and largest city.

Etymology

Victoria, like Queensland, was named after Queen Victoria, the monarch at the time.

History

After the founding of the colony of New South Wales in 1788, Australia was divided into an eastern half named New South Wales and a western half named New Holland, under the administration of the colonial government in Sydney. The first European settlement in the area later known as Victoria was established in October 1803 under Lieutenant-Governor David Collins at Sullivan Bay, Victoria on Port Phillip. It consisted of 308 convicts, 51 marines, 17 free settlers, 12 civil officers, a missionary and his wife. They had been sent from England in HMS *Calcutta* under the command of Captain Daniel Woodriff, principally out of fear that the French, who had been exploring the area, might establish their own settlement and thereby challenge British rights to the continent.

Victoria's next settlement was at Portland, on the west coast of what is now Victoria. Edward Henty settled Portland Bay in 1834.[6] Shortly after arriving from Tasmania in December 1834, Henty (aged 24 yrs) set off for thickly-populated country inland from Portland, with resident whalers, an Aboriginal woman abducted by one of them and 14 hunting dogs (a cross between greyhound and wolfhound}. On spotting a single Aboriginal man, the men set the dogs on him.[7]

Melbourne was founded in 1835 by John Batman. From settlement the region around Melbourne was known as the Port Phillip District, a separately administered part of New South Wales.

Impact of colonisation on Aborigines

At the start of colonisation, what is now Victoria, was already settled and well populated by Aboriginal groups. The problem was that this was not recognised—legally or in practice—by the British Crown. With pastoral land in the colony of Van Diemen's Land fully allocated to colonists, and Batman himself already a significant land-owner, he turned his attention to mainland land speculation at Port Phillip Bay, which began in 1835 without the consent of the British Crown and at an enormous cost in the lives and livelihoods of its Aboriginal land-owners. With no legal recognition or protection of the Aboriginal land-owners, violence ensued. For example, in August 1836, Aborigines killed the squatter Charles Franks and an unnamed shepherd, at Franks' station on the Werribee River (near Melbourne). In response, Henry Batman (John Batman's brother) led an indiscriminate punitive expedition against 70–80 Aborigines (men, women and children) living in nine large huts on the Werribee River, killing an unrecorded number.[8] In spite of this, in May 1837, Henry Batman ". . . was appointed acting Commissioner of Crown Lands, the official charged with overseeing the squatters".[9] Earlier, on 4 March 1837, Governor Bourke in his visit to Melbourne addressed 120 Aborigines, ". . . whom he exhorted . . . to good conduct and attention to the Missionary". The Kulin were given blankets and four favoured men, who had been recommended for "honorary distinctions" by [Police Magistrate Captain William] Lonsdale, were awarded brass plates.[10]

By 1839, large numbers of homeless, dispossessed Aborigines, refugees from surrounding pastoral districts, were ". . . surviving whenever and however they could on the geographic, social and economic margins of the town [i.e. Melbourne]." [11] When George Augustus Robinson arrived in the town in the winter of 1839, " . . . four to five hundred blacks of the Port Phillip tribes . . ." were gathering at a camp site on the south bank of the Yarra River, suffering hunger and disease.[12] By 1840, Robinson still ". . . had no stores allocated to him . . ." by Captain William Lonsdale (colonist), the Police Magistrate in Melbourne, even though ". . . it was patently obvious that the Aborigines were starving, and many were ill and near death . . .".[13] With land in the hinterland overrun by ". . . vast numbers of sheep and cattle . . ." and ". . . conditions in the countryside becoming intolerable, the blacks swarmed into Melbourne looking for food and blankets".[14]

Between 1836 and 1842, Victorian Aboriginal groups were largely dispossessed of territory bigger than England. [15] By January 1844, there were said to be 675 Aborigines resident in squalid camps in Melbourne. [16] Although the British Colonial Office appointed 5 "Aboriginal Protectors" for the entire Aboriginal population of Victoria, arriving in Melbourne in 1839, they worked ". . . within a land policy that nullified their work, and there was no political will to change this".[17] "It was government policy to encourage squatters to take possession of whatever [Aboriginal] land they chose, . . . that largely explains why almost all the original inhabitants of Port Phillip's vast grasslands were dead so soon after 1835".[18] By 1845, fewer than 240 wealthy Europeans held all the pastoral licences then issued in Victoria and became the patriarchs " . . . that were to wield so much political and economic power in Victoria for generations to come".[19]

Creation of separate colony of Victoria

With the Aboriginal population dispossessed of their lands and their management of fire having been disrupted for almost 15 years, the Colony experienced for the first time its largest ever bushfires, burning about 25% of the land area of Victoria on Black Thursday (1851) on 6 February 1851.

On 15 November 1851, the British Government separated the area from New South Wales, proclaiming a new Colony of Victoria. Days later, still in 1851 gold was discovered near Ballarat, and subsequently at Bendigo. Later discoveries occurred at many sites across Victoria. This triggered one of the largest gold rushes the world has ever seen. The colony grew rapidly in both population and economic power. In ten years the population of Victoria increased sevenfold from 76,000 to 540,000. All sorts of gold records were produced including the "richest shallow alluvial goldfield in the world" and the largest gold nugget. Victoria produced in the decade 1851–1860 20 million ounces of gold, one third of the world's output.

Immigrants arrived from all over the world to search for gold, especially from Ireland and China. Many Chinese miners worked in Victoria, and their legacy is particularly strong in Bendigo and its environs. Although there was some racism directed at them, there was not the level of anti-Chinese violence that was seen at the Lambing Flat riots in New South Wales. However, there was a riot at Buckland Valley near Bright in 1857. Conditions on the gold fields were cramped and unsanitary; an outbreak of typhoid at Buckland Valley in 1854 killed over 1,000 miners.

Swearing Allegiance to the Southern Cross at the Eureka Stockade on 1 December 1854 – watercolour by Charles Doudiet

In 1854 at Ballarat there was an armed rebellion against the government of Victoria by miners protesting against mining taxes (the "Eureka Stockade"). This was crushed by British troops, but the discontents prompted colonial authorities to reform the administration (particularly reducing the hated mining licence fees) and extend the franchise. Within a short time, the Imperial Parliament granted Victoria responsible government with the passage of the Colony of Victoria Act 1855. Some of the leaders of the Eureka rebellion went on to became members of the Victorian Parliament.

The first foreign military action by the colony of Victoria was to send troops and a warship to New Zealand as part of the Māori Wars. Troops from New South Wales had previously participated in the Crimean War.

In 1901 Victoria became a state in the Commonwealth of Australia. As a result of the gold rush, Melbourne had by then become the financial centre of Australia and New Zealand. Between 1901 and 1927, Melbourne was the capital of Australia while Canberra was under construction. It was also the largest city in Australia at the time.

On Saturday 7 February 2009 ("Black Saturday"), the state was affected by the 2009 Victorian bushfires, resulting in 173 deaths.[20] [21]

Government

The Victorian Parliament House, built in 1856, stands in Spring Street, Melbourne.

The Legislative Council Chamber, as photographed in 1878

One of many local government seats, Geelong Town Hall

Composition of the Parliament of Victoria		
Political Party	**Legislative Assembly**	**Legislative Council**
ALP	43	16
Liberal	35	18
National	10	3
Greens	0	3
Source: Victorian Electoral Commission		

Parliament

Victoria has a parliamentary form of government based on the Westminster System. Legislative power resides in the Parliament consisting of the Governor (the representative of the Queen), the executive (the Government), and two legislative chambers. The Parliament of Victoria consists of the lower house Legislative Assembly, the upper house Legislative Council and the Queen of Australia.

Eighty-eight members of the Legislative Assembly are elected to four-year terms from single-member electorates.

In November 2006, the Victorian Legislative Council elections were held under a new multi-proportional representation system. The State of Victoria was divided into eight electorates with each electorate represented by five representatives elected by Single Transferable Vote proportional representation. The total number of upper house members was reduced from 44 to 40 and their term of office is now the same as the lower house members—four years. Elections for the Victorian Parliament are now fixed and occur in November every four years. Prior to the 2006 election, the Legislative Council consisted of 44 members elected to eight-year terms from 22 two-member electorates.

Premier and cabinet

The Premier of Victoria is the leader of the political party or coalition with the most seats in the Legislative Assembly. The Premier is the public face of government and, with cabinet, sets the legislative and political agenda. Cabinet consists of representatives elected to either house of parliament. It is responsible for managing areas of government that are not exclusively the Commonwealth's, by the Australian Constitution, such as education, health and law enforcement. The current Premier of Victoria is Ted Baillieu.

Governor

Executive authority is vested in the Governor of Victoria who represents and is appointed by Queen Elizabeth II. The post is usually filled by a retired prominent Victorian. The governor acts on the advice of the premier and cabinet. The current Governor of Victoria is Alex Chernov.

Constitution

Victoria has a written constitution [22]. Enacted in 1975, but based on the 1855 colonial constitution [23], it establishes the parliament as the state's law-making body for matters coming under state responsibility. The Victorian Constitution can be amended by the parliament of Victoria. Under new provisions to be enacted, changes to the Victorian Constitution will be subjected to a plebiscite of votes, voting in a referendum.

Politics

Premier Ted Baillieu leads a Liberal/National Coalition that won the November 2010 Victorian state election.

The centre-left Australian Labor Party (ALP), the centre-right Liberal Party of Australia and the rural-based National Party of Australia are Victoria's major political parties. Traditionally, Labor is strongest in Melbourne's inner, working class and western and northern suburbs, Morwell, Ballarat, Bendigo and Geelong. The Liberals' main support lies in Melbourne's more affluent eastern and outer suburbs, and some rural and regional centres. The Nationals are strongest in Victoria's North Western and Eastern rural regional areas.

Federal government

Victorian voters elect 49 representatives to the Parliament of Australia, including 37 members of the House of Representatives and 12 members of the Senate. Since 2010, the ALP has held 22 Victorian house seats, the Liberals 12, the Nationals two and the Greens one. As of 1 July 2008, the Liberals have held six senate seats, the ALP five and the Family First Party one.

Local government

Victoria is incorporated into 79 municipalities for the purposes of local government, including 39 shires, 32 cities, seven rural cities and one borough. Shire and city councils are responsible for functions delegated by the Victorian parliament, such as city planning, road infrastructure and waste management. Council revenue comes mostly from property taxes and government grants.

Source: Victorian Parliamentary Library, Department of Victorian Communities, Australian Electoral Commission

Demographics

The 2006 Australian census reported that Victoria had 4,932,422 people resident at the time of the census, an increase of 6.2% on the 1996 figure. The Australian Bureau of Statistics estimates that in June 2010 the state's population reached 5,547,500, an increase of 1.8% from the previous year and may well reach 7.2 million by 2050.

Melbourne, the state capital, is home to more than seven in ten Victorians.

Victoria's founding Anglo-Celtic population has been supplemented by successive waves of migrants from southern and eastern Europe, Southeast Asia and, most recently, the Horn of Africa and the Middle East. Victoria's population is ageing in proportion with the average of the remainder of the Australian population.

About 72% of Victorians are Australian-born. This figure falls to around 66% in Melbourne but rises to higher than 95% in some rural areas in the north west of the state. Around two-thirds of Victorians claim Australian, Scottish, English or Irish ancestry. Less than 1% of Victorians identify themselves as Aboriginal. The largest groups of people born outside Australia came from the British Isles, China, Italy, Vietnam, Greece and New Zealand.

More than 70% of Victorians live in Melbourne, located in the state's south. The greater Melbourne metropolitan area is home to an estimated 3.9 million people.[24] Leading urban centres include Geelong, Ballarat, Bendigo, Shepparton, Mildura, Warrnambool, Wodonga and the Latrobe Valley.

Population growth estimates for Victoria	
2007	5,087,000
2011	5,500,000
2016	6,000,000
2021	6,400,000
2026	6,800,000
2031	7,300,000
Source: Dept of Planning and Community Development	

Victoria is Australia's most urbanised state: nearly 90% of residents living in cities and towns. State Government efforts to decentralise population have included an official campaign run since 2003 to encourage Victorians to settle in regional areas,[25] however Melbourne continues to rapidly outpace these areas in terms of population growth.[26]

Age structure and fertility

The government predicts that nearly a quarter of Victorians will be aged over 60 by 2021. The 2006 census reveals that Australian average age has crept upward from 35 to 37 since 2001, which reflects the population growth peak of 1969–72. In 2007, Victoria recorded a TFR of 1.87, the highest after 1978. [27]

Crime

The state of Victoria is divided into four geographical regions; North-West Metropolitan Region, Southern Metropolitan Region, Eastern Region, Western Region. In 2010-2011 there were 152 homicides within the state of Victoria.[28]

Religion

About 60.5% of Victorians describe themselves as Christian. Roman Catholics form the single largest religious group in the state with 27.5% of Victorian population, followed by Anglicans and members of the Uniting Church. Catholics and Protestants (including Anglicans) in Victoria each form around 30% of the population. Buddhism, the state's largest non-Christian religion, is also the fastest growing with 132,634. Victoria is also home of 109,370 Muslims and 41,105 Jews. Around 20% of Victorians claim no religion, and even amongst those who declare a religious affiliation, church attendance is low.[29]

In 2008, the levels of couples choosing to marry in a church had dropped to 36%; the other 64% chose to register their marriage with a civil celebrant.[30]

Education

Primary and secondary

Victoria's state school system dates back to 1872, when the colonial government legislated to make schooling both free and compulsory. The state's public secondary school system began in 1905. Before then, only private secondary schooling was available. Today, a Victorian school education consists of seven years of primary schooling (including one preparatory year) and six years of secondary schooling.

The final years of secondary school are optional for children aged over 17. Victorian children generally begin school at age five or six. On completing secondary school, students earn the Victorian Certificate of Education. Students who successfully complete their secondary education also receive a tertiary entrance ranking, or ATAR score, to determine university admittance.

Victorian schools are either publicly or privately funded. Public schools, also known as state or government schools, are funded and run directly by the Victoria Department of Education [31]. Students do not pay tuition fees, but some extra costs are levied. Private fee-paying schools include parish schools run by the Roman Catholic Church and independent schools similar to English public schools. Independent schools are usually affiliated with Protestant churches. Victoria also has several private Jewish and Islamic primary and secondary schools. Private schools also receive some public funding. All schools must comply with government-set curriculum standards. In addition, Victoria has four government selective schools, Melbourne High School for boys, MacRobertson Girls' High School for girls, the coeducational schools John Monash Science School, Nossal High School and Suzanne Cory High School, and The Victorian College of the Arts Secondary School. Students at these schools are exclusively admitted on the basis of a selective entry test.

As of August 2010, Victoria had 1,548 public schools, 489 Catholic schools and 214 independent schools. Just under 540,800 students were enrolled in public schools, and just over 311,800 in private schools. Over 61 per cent of private students attend Catholic schools. More than 462,000 students were enrolled in primary schools and more than 390,000 in secondary schools. Retention rates for the final two years of secondary school were 77 per cent for public school students and 90 per cent for private school students. Victoria has about 63,519 full-time teachers.[32]

Tertiary education

Victoria has nine universities. The first to offer degrees, the University of Melbourne, enrolled its first student in 1855. The largest, Monash University, has an enrolment of nearly 56,000 students—more than any other Australian university. Both the University of Melbourne and Monash University are purportedly ranked highly among the world's best universities requiring a fairly high entry score, passing of mature age entrance exams or direct payment for student admission into their courses.

The University of Melbourne is Victoria's oldest university.

The number of students enrolled in Victorian universities was 241,755 at 2004, an increase of 2% on the previous year. International students made up 30% of enrolments and account for the highest percentage of pre-paid university tuition fees. The largest number of enrolments were recorded in the fields of business, administration and economics, with nearly a third of all students, followed by arts, humanities, and social science, with 20% of enrolments.

Victoria has 18 government-run institutions of "technical and further education" (TAFE). The first vocational institution in the state was the Melbourne Mechanics' Institute (established in 1839), which is now the Melbourne

Athenaeum. More than 1,000 adult education organisations are registered to provide recognised TAFE programs. In 2004, there were about 480,700 students enrolled in vocational education programs in the state.

Source: Australian Bureau of Statistics, Department of Education and Training (Victoria), Department of Education, Science and Training (Commonwealth), National Centre for Vocational Education Research

Libraries

The State Library of Victoria is the State's research and reference library. It is responsible for collecting and preserving Victoria's documentary heritage and making it available through a range of services and programs. Material in the collection includes books, newspapers, magazines, journals, manuscripts, maps, pictures, objects, sound and video recordings and databases.

The State Library of Victoria forecourt

In addition, local governments maintain local lending libraries, typically with multiple branches in their respective municipal areas.

Economy

Victorian production and workers by economic activities			
Economic sector	**GSP produced**	**Number of workers**	**Percentage of workers**
Finance, insurance and property	30.5%	319,109	15.3%
Community, social and personal services	16.6%	562,783	27.4%
Manufacturing	15.4%	318,218	15.3%
Wholesale and retail trade	12.1%	423,328	20.3%
Transport, utilities and communications	10.6%	133,752	6.4%
Construction	6.2%	136,454	6.6%
Government	4%	62,253	3%
Agriculture	3.3%	72,639	3.5%
Mining	1.3%	4,472	0.2%
Other	–	49,208	2%
Source: Australian Bureau of Statistics. Figures are for 2004–2005			

The state of Victoria is the second largest economy in Australia after New South Wales, accounting for a quarter of the nation's gross domestic product. The total gross state product (GSP) at current prices for Victoria was at just over A$293 billion, with a GSP per capita of A$52,872. The economy grew by 2.0 per cent in 2010, less than the Australian average of 2.3 per cent.

Finance, insurance and property services form Victoria's largest income producing sector, while the community, social and personal services sector is the state's biggest employer. Despite the shift towards service industries, the troubled manufacturing sector remains Victoria's single largest employer and income producer. As a result of job losses in declining sectors such as manufacturing, Victoria has the highest unemployment rate in Australia as of September 2009.[33]

Agriculture

Victoria's stand at the Paris *Exhibition Universal* of 1867, showing bales of wool

During 2003–04, the gross value of Victorian agricultural production increased by 17% to $8.7 billion. This represented 24% of national agricultural production total gross value. As of 2004, an estimated 32,463 farms occupied around 136,000 square kilometres (52,500 sq mi) of Victorian land. This comprises more than 60% of the state's total land surface. Victorian farms range from small horticultural outfits to large-scale livestock and grain productions. A quarter of farmland is used to grow consumable crops.

More than 26,000 square kilometres (10,000 sq mi) of Victorian farmland are sown for grain, mostly in the state's west. More than 50% of this area is sown for wheat, 33% for barley and 7% for oats. A further 6,000 square kilometres (2,300 sq mi) is sown for hay. In 2003–04, Victorian farmers produced more than 3 million tonnes of wheat and 2 million tonnes of barley. Victorian farms produce nearly 90% of Australian pears and third of apples. It is also a leader in stone fruit production. The main vegetable crops include asparagus, broccoli, carrots, potatoes and tomatoes. Last year, 121,200 tonnes of pears and 270,000 tonnes of tomatoes were produced.

More than 14 million sheep and 5 million lambs graze over 10% of Victorian farms, mostly in the state's north and west. In 2004, nearly 10 million lambs and sheep were slaughtered for local consumption and export. Victoria also exports live sheep to the Middle East for meat and to the rest of the world for breeding. More than 108,000 tonnes of wool clip was also produced—one-fifth of the Australian total.

Victoria is the centre of dairy farming in Australia. It is home to 60% of Australia's 3 million dairy cattle and produces nearly two-thirds of the nation's milk, almost 6.4 million litres. The state also has 2.4 million beef cattle, with more than 2.2 million cattle and calves slaughtered each year. In 2003–04, Victorian commercial fishing crews and aquaculture industry produced 11,634 tonnes of seafood valued at nearly A$109 million. Blacklipped abalone is the mainstay of the catch, bringing in A$46 million, followed by southern rock lobster worth A$13.7 million. Most abalone and rock lobster is exported to Asia.

Manufacturing

Machinery and equipment manufacturing is the state's most valuable manufacturing activity, followed by food and beverage manufacturing and petroleum, coal and chemical manufacturing. More than 15% Victorian workers are employed in manufacturing industries. Victoria has 318,000 manufacturing workers. The state is marginally behind New South Wales in the value of manufacturing output.

Major industrial plants belong to the car manufacturers Ford, Toyota and Holden; Alcoa's Portland and Point Henry aluminium smelters; oil refineries at Geelong and Altona; and a major petrochemical facility at Laverton.

Victoria also plays an important role in providing goods for the defence industry. Melbourne is the centre of manufacturing in Victoria, followed by Geelong. Energy production has aided industrial growth in the Latrobe Valley.

Mining

Mining in Victoria contributes around A$3 billion to the gross state product (~1%) but employs less than 1% of workers. The Victorian mining industry is concentrated on energy producing minerals, with brown coal, petroleum and gas accounting for nearly 90% of local production. The oil and gas industries are centred off the coast of Gippsland in the state's east, while brown coal mining and power generation is based in the Latrobe Valley.

Yallourn Power Station in the Latrobe Valley

In the 2005/2006 fiscal year, the average gas production was over 700 million cubic feet (m^3) per day (M cuft/d) and represented 18% of the total national gas sales, with demand growing at 2% per year.[34]

In 1985, oil production from the offshore Gippsland Basin peaked to an annual average of 450000 barrels (72000 m^3) per day. In 2005–2006, the average daily oil production declined to 83000 bbl (13200 m^3)/d, but despite the decline Victoria still produces almost 19.5% of crude oil in Australia.[35]

Brown coal is Victoria's leading mineral, with 66 million tonnes mined each year for electricity generation in the Latrobe Valley, Gippsland.[36] The region is home to the world's largest known reserves of brown coal.

Despite being the historic centre of Australia's gold rush, Victoria today contributes a mere 1% of national gold production. Victoria also produces limited amounts of gypsum and kaolin.

Service industry

The service industries sector is the fastest growing component of the Victorian economy. It includes the wide range of activities generally classified as community, social and personal services; finances, insurance and property services, government services, transportation and communication, and wholesale and retail trade. Most service industries are located in Melbourne and the state's larger regional centres.

As of 2004–05, service industries employed nearly three-quarters of Victorian workers and generated three-quarters of the state's GSP. Finance, insurance and property services, as a group, provide a larger share of GSP than any other economic activity in Victoria. More than a quarter of Victorian workers are employed by the community, social and personal services sector.[37]

Geology and geography

Victoria's northern border is the southern bank of the Murray River. It also rests at the southern end of the Great Dividing Range, which stretches along the east coast and terminates west of Ballarat. It is bordered by South Australia to the west and shares Australian's shortest land border with Tasmania. The official border between Victoria and Tasmania is at 39°12' S, which passes through Boundary Islet in the Bass Strait for 85 metres.[38] [39] Victoria contains many topographically, geologically and climatically diverse areas, ranging from the wet, temperate

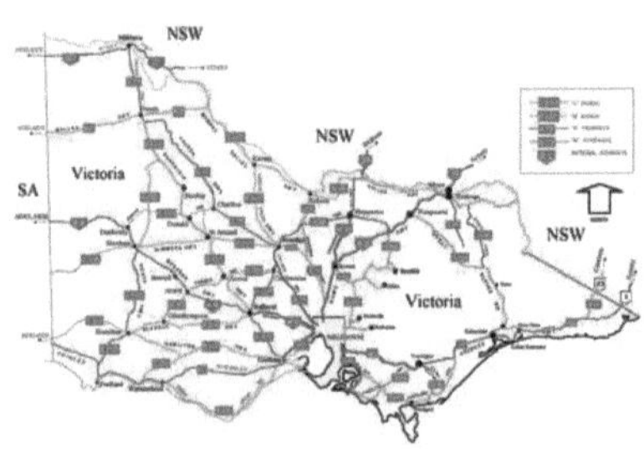

Victorian cities, towns, settlements and road network

climate of Gippsland in the southeast to the snow-covered Victorian alpine areas which rise to almost 2,000 metres (6,500 ft), with Mount Bogong the highest peak at 1,986 m; (6,516 ft). There are extensive semi-arid plains to the west and northwest.

There is an extensive series of river systems in Victoria. Most notable is the Murray River system. Other rivers include: Ovens River, Goulburn River, Patterson River, King River, Campaspe River, Loddon River, Wimmera River, Elgin River, Barwon River, Thomson River, Snowy River, Latrobe River, Yarra River, Maribyrnong River, Mitta River, Hopkins River, Merri River and Kiewa River.

The state symbols include the Pink Heath (state flower), Leadbeater's Possum (state animal) and the Helmeted Honeyeater (state bird).

The state's capital, Melbourne, contains approximately 70% of the state's population and dominates its economy, media, and culture. For other cities and towns, see List of localities (Victoria) and Local Government Areas of Victoria.

Transport

Victoria has the highest population density in any state in Australia, with population centres spread out over most of the state; only the far northwest and the Victorian Alps lack permanent settlement.

The Victorian road network services the population centres, with highways generally radiating from Melbourne and other major cities and rural centres with secondary roads interconnecting the highways to each other. Many of the highways are built to freeway standard ("M" freeways), while most are generally sealed and of reasonable quality.

A current Melbourne C class (Citadis) tram

Rail transport in Victoria is provided by several private and public railway operators who operate over government-owned lines. Major operators include: Metro Trains Melbourne which runs an extensive, electrified, passenger system throughout Melbourne and suburbs; V/Line which is now owned by the Victorian Government, operates a concentrated service to major regional centres, as well as long distance services on other lines; Pacific National, CFCLA, El Zorro which operate freight services; Great Southern Railway which operates The Overland Melbourne—Adelaide; and CountryLink which operates XPTs Melbourne—Sydney.

There are also several smaller freight operators and numerous tourist railways operating over lines which were once parts of a state-owned system. Victorian lines mainly use the 5 ft 3 in (1600 mm) broad gauge. However, the interstate trunk routes, as well as a number of branch lines in the west of the state have been converted to 4 ft 8 $^{1}/_{2}$ in (1435 mm) standard gauge. Two tourist railways operate over 2 ft 6 in (762 mm) narrow gauge lines, which are the remnants of five formerly government-owned lines which were built in mountainous areas.

Melbourne has the world's largest tram network,[40] currently operated by Yarra Trams. As well as being a popular form of public transport, over the last few decades trams have become one of Melbourne's major tourist attractions. There are also tourist trams operating over portions of the former Ballarat and Bendigo systems. There are also tramway museums at Bylands and Haddon.

Melbourne Airport is the major domestic and international gateway for the state. Avalon Airport is the state's second busiest airport, which is complements Essendon and Moorabbin Airports to see the remainder of Melbourne's air traffic. Hamilton Airport, Mildura Airport, Mount Hotham and Portland Airport are the remaining airports with scheduled domestic flights. There are no fewer than 27 other airports in the state with no scheduled flights.

The Port of Melbourne is the largest port for containerised and general cargo in Australia,[41] and is located in Melbourne on the mouth of the Yarra River, which is at the head of Port Phillip. Additional seaports are at

Westernport, Geelong, and Portland.

CountryLink is an interstate train and coach service provider which provides twice-daily services from Melbourne to Sydney.

An 'X'Trapolis' in livery of former commuter-service provider Connex, now succeeded by Metro Trains Melbourne

V/Line is a government-owned train and coach service provider in Victoria. The enterprise provides inter-city services to a number of regional cities in the state.

Utilities

Energy

Victoria's major utilities include a collection of brown-coal-fired power stations, particularly in the Latrobe Valley. One of these is Hazelwood Power Station, which is number 1 in the worldwide List of least carbon efficient power stations.

Water

Victoria's water infrastructure includes a series of dams and reservoirs, predominantly in Central Victoria, that hold and collect water for much of the state. The water collected is of a very high quality and requires little chlorination treatment, giving the water a taste more like water collected in a rainwater tank. In regional areas however, such as in the west of the state, chlorination levels are much higher.

The Victorian Water Grid consists of a number of new connections and pipelines being built across the State. This allows water to be moved around Victoria to where it is needed most and reduces the impact of localised droughts in an era thought to be influenced by climate change. Major projects already completed as part of the Grid include the Wimmera Mallee Pipeline and the Goldfields Superpipe.[42]

Climate

Average monthly maximum temperature in Victoria		
Month	**Melbourne**	**Mildura**
January	25.8 °C	32.8 °C
February	25.8 °C	32.7 °C
March	23.8 °C	29.3 °C
April	20.2 °C	24.1 °C
May	16.6 °C	19.6 °C
June	14.0 °C	16.0 °C
July	13.4 °C	15.4 °C
August	14.9 °C	17.7 °C
September	17.2 °C	21.1 °C
October	19.6 °C	25.0 °C
November	21.8 °C	29.0 °C
December	24.1 °C	31.7 °C
Source: Bureau of Meteorology		

Victoria has a varied climate despite its small size. It ranges from semi-arid and hot in the north-west, to temperate and cool along the coast. Victoria's main land feature, the Great Dividing Range, produces a cooler, mountain climate in the centre of the state.

Victoria's southernmost position on the Australian mainland means it is cooler and wetter than other mainland states and territories. The coastal plain south of the Great Dividing Range has Victoria's mildest climate. Air from the Southern Ocean helps reduce the heat of summer and the cold of winter. Melbourne and other large cities are located in this temperate region.

The Mallee and upper Wimmera are Victoria's warmest regions with hot winds blowing from nearby deserts. Average temperatures top 30 °C (86 °F) during summer and 15 °C (59 °F) in winter. Victoria's highest maximum temperature of 48.8 °C (119.9 °F) was recorded in Hopetoun on 7 February 2009, during the 2009 southeastern Australia heat wave.[43]

The Victorian Alps in the northeast are the coldest part of Victoria. The Alps are part of the Great Dividing Range mountain system extending east-west through the centre of Victoria. Average temperatures are less than 9 °C (48 °F) in winter and below 0 °C (32 °F) in the highest parts of the ranges. The state's lowest minimum temperature of −11.7 °C (10.9 °F) was recorded at Omeo on 13 June 1965, and again at Falls Creek on 3 July 1970.[43]

Rainfall

Victoria is the wettest Australian state after Tasmania. Rainfall in Victoria increases from north to south, with higher averages in areas of high altitude. Median annual rainfall exceeds 1,800 millimetres (71 in) in some parts of the northeast but is less than 250 millimetres (10 in) in the Mallee.

Rain is heaviest in the Otway Ranges and Gippsland in southern Victoria and in the mountainous northeast. Snow generally falls only in the mountains and hills in the centre of the state. Rain falls most frequently in winter, but summer precipitation is heavier. Rainfall is most reliable in Gippsland and the Western District, making them both leading farming areas. Victoria's highest recorded daily rainfall was 375 millimetres (14.7 in) at Tanybryn in the Otway Ranges on 22 March 1983.[43]

Average temperatures and precipitation for Victoria

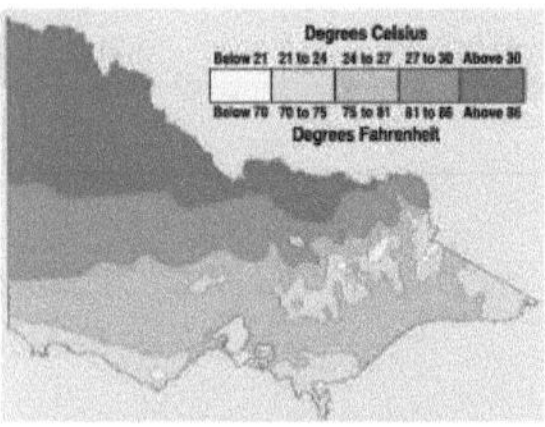

Average January temperatures: Victoria's north is almost always hotter than coastal and mountainous areas.

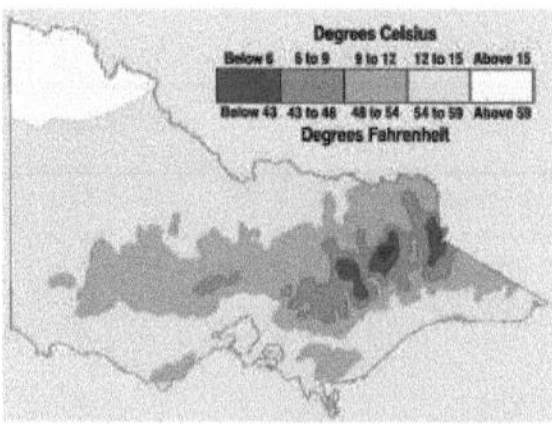

Average July temperatures: Victoria's hills and ranges are coolest during winter. Snow also falls there.

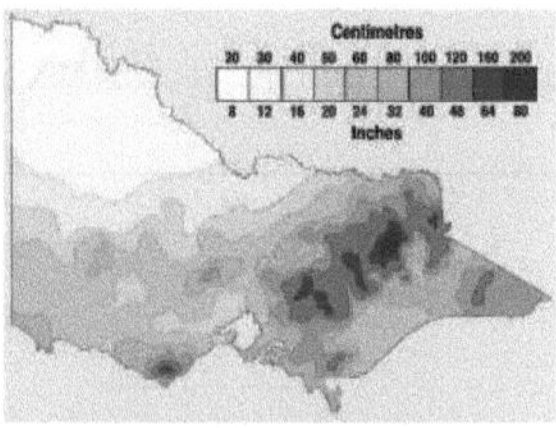

Average yearly precipitation: Victoria's rainfall is concentrated in the mountainous north-east and coast.

Source: Bureau of Meteorology, Department of Primary Industries, Australian Natural Resources Atlas

Tourism

Some major tourist destinations in Victoria are:

The Twelve Apostles

- The metropolis of Melbourne, particular its inner city suburbs (known also for shopping tourism) and the attractions of the city centre such as Crown Casino, Melbourne Zoo, Melbourne Museum, the Melbourne Aquarium, ScienceWorks, Healesville Sanctuary, Werribee Open Range Zoo, tourism precincts such as Melbourne Docklands, Southbank and St Kilda as well as cultural and sporting tourist icons such as The Arts Centre, National Gallery of Victoria, the Melbourne Cricket Ground, also known as the MCG, and the Eureka Tower, tallest building in the Southern Hemisphere, with its Skydeck 88.
- Victoria has more than 2000 kilometres of coastline with hundreds of beaches.[44]
- The Goldfields region featuring the historic cities of Ballarat, Beechworth, Bendigo, Castlemaine, Maldon and Daylesford.
- Natural attractions, such as The Twelve Apostles, Wilsons Promontory, The Grampians, the Fairy Penguins (particularly at Phillip Island and St Kilda), the Buchan Caves and the Gippsland Lakes.
- The Dandenong Ranges (in particular the Puffing Billy Railway).
- Towns along the Murray River and Riverina including Echuca and Mildura including waterskiing.
- Geelong and the Australian International Airshow
- The Bellarine Peninsula which features historic resort towns such as Queenscliff.
- The Surf Coast which features famous beaches such as Bells Beach, Torquay and Lorne
- Mornington Peninsula, particularly for its wineries and secluded beaches, Arthur's Seat and the coastal attractions of Portsea and Sorrento.
- Yarra Valley (in particular Healesville Sanctuary and wineries).
- Great Ocean Road, which features The Twelve Apostles, historic towns of Port Fairy and Portland, cliffs and whale watching and resort towns such as Lorne.
- The Victorian Alpine Region, part of the Australian Alps, particularly for skiing
- The Central Victorian Highlands, 'Highcountry' are very well known for winter sports and bushwalking
- Wine regions across the entire state.

Other popular tourism activities are gliding, hang-gliding, hot air ballooning and scuba diving.

Major events also play a big part in tourism in Victoria, particularly cultural tourism and sports tourism. Most of these events are centred around Melbourne, but others occur in regional cities, such as the V8 Supercars and Australian Motorcycle Grand Prix at Phillip Island, the Grand Annual Steeplechase at Warrnambool and the Australian International Airshow at Geelong and numerous local festivals such as the popular Port Fairy Folk Festival, Queenscliff Music Festival, Bells Beach SurfClassic and the Bright Autumn Festival.

Sport

The Melbourne Cricket Ground during the 1998 Boxing Day Test match

Victoria is the home of Australian rules football, with ten of the seventeen clubs of the Australian Football League based in Victoria, and the traditional Grand Final held at the Melbourne Cricket Ground usually on the last Saturday in September.

Victoria's cricket team, the Victorian Bushrangers play in the national Sheffield Shield cricket competition. Victoria is represented in the National Rugby League by the Melbourne Storm and in Super Rugby by the Melbourne Rebels. It is also represented in Football (soccer) by Melbourne Victory and Melbourne Heart in the A-League.

Melbourne has held the 1956 Summer Olympics, 2006 Commonwealth Games and the FINA World Swimming Championship.

Melbourne is also home to the Australian Open tennis tournament in January each year, the first of the world's four Grand Slam tennis tournaments, and the Australian Formula One Grand Prix.

Victoria's Bells Beach hosts one of the world's longest-running surfing competition, the Bells Beach SurfClassic, which is part of The ASP World Tour.

Netball is a big part of sport in Victoria. The Melbourne Vixens represent Victoria in the ANZ Championship. Some of the worlds best netballers such as Sharelle McMahon, Renae Hallinan, Madison Browne, Julie Corletto and Bianca Chatfield come from Victoria.

Possibly Victoria's most famous island, Phillip Island, is home of the Phillip Island Grand Prix Circuit which hosts the Australian motorcycle Grand Prix which features MotoGP (the world's premier motorcycling class), as well as the Australian round of the World Superbike Championship and the domestic V8 Supercar racing, which also visits Sandown Raceway and the rural Winton Motor Raceway circuit.

Australia's most prestigious footrace, the Stawell Gift, is an annual event.

Victoria is also home to the Aussie Millions poker tournament, the richest in the Southern hemisphere.

The Melbourne Spring Racing Carnival is one of the biggest horse racing events in the world and is one of the world's largest sporting events. The main race is for the $6 million Melbourne Cup, and crowds for the carnival exceed 700,000.

See also

- 2003 Melbourne thunderstorm
- Atlas of Victorian Birds

Geography:

- Goombala Road
- Victorian Alps
- Vicmap Topographic Map Series

Lists:

- List of highways in Victoria
- List of localities (Victoria)
- Protected areas of Victoria (Australia)

References

[1] 3101.0 – Australian Demographic Statistics, June 2010 (http://www.abs.gov.au/ausstats/abs@.nsf/mf/3101.0/), Australian Bureau of Statistics

[2] 5220.0 – Australian National Accounts: State Accounts, 2009–10 (http://www.abs.gov.au/AUSSTATS/abs@.nsf/Lookup/5220.0Main+Features12009-10?OpenDocument).

[3] "Floral Emblem of Victoria" (http://www.anbg.gov.au/emblems/vic.emblem.html). www.anbg.gov.auhi.. Retrieved 26 March 2008.

[4] "Victoria" (http://www.parliament.curriculum.edu.au/vic.php3). *Parliament@Work*.. Retrieved 27 February 2008.

[5] http://www.vic.gov.au//

[6] James Boyce (2011) 1835: The Founding of Mebourne and the Conquest of Australia, Black Inc, p.12

[7] James Boyce (2011) 1835: The Founding of Mebourne and the Conquest of Australia, Black Inc, p.14

[8] James Boyce, 1835: The Founding of Melbourne and the Conquest of Australia, Black Inc, 2011, pp.105–109

[9] James Boyce, (2008) 1835: The Founding of Melbourne and the Conquest of Australia, Black Inc., p.149

[10] James Boyce, (2008) 1835: The Founding of Melbourne and the Conquest of Australia, Black Inc., p.148

[11] James Boyce, 1835: The Founding of Melbourne and the Conquest of Australia, Black Inc. p.186

[12] Vivienne Rae-Ellis (1988) Black Robinson: Protector of Aborigines, University of Melbourne Press, Melbourne, p.172

[13] Vivienne Rae-Ellis (1988) Black Robinson: Protector of Aborigines, University of Melbourne Press, Melbourne, p.178

[14] Vivienne Rae-Ellis (1988) Black Robinson: Protector of Aborigines, University of Melbourne Press, Melbourne, p.179

[15] James Boyce, 1835: The Founding of Melbourne and the Conquest of Australia, Black Inc, 2011, page 151 citing Richard Broome, "Victoria" in McGrath (ed.), Contested Ground: 129

[16] James Boyce, 1835: The Founding of Melbourne and the Conquest of Australia, Black Inc, 2011, p.186

[17] James Boyce, 1835: The Founding of Melbourne and the Conquest of Australia, Black Inc, 2011, p.177

[18] James Boyce, 1835: The Founding of Melbourne and the Conquest of Australia, Black Inc, 2011, p.199

[19] James Boyce, 1835: The Founding of Melbourne and the Conquest of Australia, Black Inc, 2011, p.163

[20] Vic bushfires death toll – Vic Police (http://www.police.vic.gov.au/content.asp?Document_ID=19190)

[21] "Victoria under siege as fires rage across state" (http://www.news.com.au/heraldsun/story/0,21985,25022560-2862,00.html). *Herald Sun* (News Limited). 2009-02-07.. Retrieved 7 February 2009.

[22] http://www.austlii.edu.au/au/legis/vic/consol_act/ca1975188/

[23] http://www.parliament.vic.gov.au/const.html

[24] "3218.0 – Regional Population Growth, Australia, 2007–08" (http://www.abs.gov.au/ausstats/abs@.nsf/Products/3218.0~2007-08~Main+Features~Victoria?OpenDocument). *ABS website*. Australian Bureau of Statistics. 23 April 2009.. Retrieved 4 November 2009.

[25] Provincial Victoria – About (http://www.provincialvictoria.vic.gov.au/about.aspx)

[26] Colebatch, Tim (24 April 2009). "Pressure grows as Melbourne rockets to 4 million" (http://www.theage.com.au/national/pressure-grows-as-melbourne-rockets-to-4-million-20090423-agt5.html). *The Age* (Australia)..

[27] 3301.0 – Births, Australia, 2009 (http://www.abs.gov.au/ausstats/abs@.nsf/mf/3301.0?OpenDocument)

[28] http://resources.news.com.au/files/2011/08/30/1226125/188901-hs-news-file-2011-crime-statistics.pdf

[29] 2006 Census Community Profile Series : Victoria (http://www.censusdata.abs.gov.au/ABSNavigation/prenav/ViewData?&action=404&documentproductno=2&documenttype=Details&tabname=Details&areacode=2&issue=2006&producttype=Community Profiles&&producttype=Community Profiles&javascript=true&textversion=false&navmapdisplayed=true&breadcrumb=PLD&&collection=Census&period=2006&producttype=Community Profiles&#Basic Community Profile)

[30] Statistics report for 2008 from the Victorian Registry of Births, Deaths and Marriages (http://www.heraldsun.com.au/news/victoria/more-victorians-ditching-the-church-and-opting-for-registry-wedding/story-e6frf7kx-1225789054971)

[31] http://www.education.vic.gov.au/

[32] 4221.0 Schools, Australia 2010 (http://www.abs.gov.au/ausstats/abs@.nsf/mf/4221.0), Australian Bureau of Statistics, 17 March 2011.

[33] Business & Finance News – Yahoo!7 Finance (http://au.biz.yahoo.com/090910/31/28ixz.html)

[34] Department of Primary Industries (http://www.dpi.vic.gov.au/dpi/nrenmp.nsf/childdocs/-56D4D5E9AEF563E84A256A800011E5D6?open)

[35] DEPARTMENT OF PRIMARY INDUSTRIES: Oil and Gas (http://www.dpi.vic.gov.au/dpi/nrenmp.nsf/childdocs/-56D4D5E9AEF563E84A256A800011E5D6?open)

[36] Australian Bureau of Statistics: Year Book Australia, 2004 – Profile of major commodities (http://abs.gov.au/ausstats/abs@.nsf/dc057c1016e548b4ca256c470025ff88/B43CD5ECE4E63E4BCA256DEA00053A0E?opendocument)

[37] Australian Bureau of Statistics, Department of Primary Industries

[38] "Victoria Tasmania border" (http://web.archive.org/web/20060102184417/http://www.ga.gov.au/education/facts/dimensions/borders.htm). Archived from the original (http://www.ga.gov.au/education/facts/dimensions/borders.htm) on 2 January 2006.. Retrieved 7 March 2006.

[39] Boundary Islet on street-directory.com.au (http://www.street-directory.com.au/sd_new/mapsearch.cgi?SuburbID=13264&star=5&PCode=3004&heading=&x=147.0211005752303&RegionID=47&y=-39.20060195855119&level=4&CouncilID=608&StateID=1)

[40] DoI (2008). (http://www.doi.vic.gov.au/DOI/DOIElect.nsf/alldocs/1173456893D78FD4CA25741A000461EE/$FILE/Investing_in_Transport_East_West-Chapter03.pdf). Retrieved 28 April 2008.

[41] "DoI media release – 'GOVERNMENT OUTLINES VISION FOR PORT OF MELBOURNE FREIGHT HUB' – 14 August 2006" (http://www.legislation.vic.gov.au/domino/Web_Notes/newmedia.nsf/798c8b072d117a01ca256c8c0019bb01/1b374a6d8cae09e9ca2571cb0003c23a!OpenDocument). . Retrieved 26 July 2007.

[42] "Department of Sustainability & Environment, "Expansion of the Water Grid", http://www.ourwater.vic.gov.au/programs/water-grid, accessed 27 January 2011".

[43] "Rainfall and Temperature Records: National" (http://www.bom.gov.au/climate/extreme/records/national.pdf) (PDF). Bureau of Meteorology. . Retrieved 14 November 2009.

[44] WA Today: Waves of fancy: Victoria's best beaches (http://www.watoday.com.au/travel/waves-of-fancy-victorias-best-beaches-20081127-6jx2.html)

Further Reading—Victorian Frontier History

- Jan Critchett, (1990), *A distant field of murder: Western district frontiers, 1834-1848*, Melbourne University Press (Carlton, Vic. and Portland, Or.) ISBN 052284389
- Ian D Clark (1990) *Aboriginal languages and clans: An historical atlas of western and central Victoria, 1800-1900*, Dept. of Geography & Environmental Science, Monash University (Melbourne), ISBN 090968541X
- Ian D Clark (1995), *Scars in the landscape: A register of massacre sites in western Victoria, 1803-1859*, Australian Institute of Aboriginal and Torres Strait Islander Studies (Canberra), ISBN 0855752815
- Ian D Clark (2003) *'That's my country belonging to me' - Aboriginal land tenure and dispossession in nineteenth century Western Victoria*, Ballarat Heritage Services, Ballarat.

External links

- State Government of Victoria website (http://www.vic.gov.au/)
- Official Government tourism website (http://www.visitvictoria.com/)
- Tourism Victoria's Online Image Library (http://www.visionsofvictoria.com/)

Dandenong_railway_station

Station information	
Code	DNG
Distance from Flinders Street	30 km
Operator	Metro Trains Melbourne
Lines	Bairnsdale, Cranbourne, Pakenham
# Platforms	3
# Tracks	3 (9 including freight)
Status	Premium station
Metlink [1] profile	Link [2]
Melway map	Link [3]
Google map	Link [4]
Metcard / Myki Zone	2

Dandenong is a railway station located in Dandenong, Victoria, Australia. Located 30 km from Flinders Street Station in central Melbourne, the station serves as the junction of the Pakenham and Cranbourne railway lines, and is the terminus station for several peak services. It also serves V/Line trains along the Traralgon and Bairnsdale lines.

Station overview

Facilities

Dandenong station is located on Foster Street, adjacent to the Dandenong central business district as well as a housing development known as Metro 3175. The station is part of Metcard Zone Zone 2, and as a Premium Station it is staffed from first train to the last each day.

The current design has a bright yellow overpass attached to three platforms, with lifts to each platform and a ramp from the street, making it wheelchair accessible. A number of train stabling sidings are near the station, as is a little used goods yard to the south of the station.

Stabling facilities are provided on the City side of the station on either side of the track, but trains can also be stabled on tracks parallel to the platforms.

History

Dandenong station was first opened in 1877 as one of the original stations on the Melbourne to Sale railway line. The South Gippsland Railway Line opened in 1892 from Dandenong. The station buildings was reconstructed in 1975,[5] and again in 1994-1995, as part of the $27 million Federal Government funded Cranbourne electrification project.[6] The new station opened in March 1995.[7]

A third rail line is to be built connecting the station with Caulfield railway station as part of the $1 billion Dandenong railway line triplication project.[8]

Platforms

Platforms 1, 2 and 3:

- Metro Trains services towards Flinders Street.
- Metro Trains services towards Cranbourne and Pakenham.
- V/Line Gippsland line services towards Southern Cross Station (not picking up passengers).
- V/Line Gippsland line services towards Warragul and Traralgon. Some services extend to Sale and Bairnsdale (not setting down passengers).

References

- Dandenong information at Vicsig.net [9]

[1] http://www.metlinkmelbourne.com.au
[2] http://www.metlinkmelbourne.com.au/stop/view/19888
[3] http://www.street-directory.com.au/sd_new/genmap.cgi?x=145.209120412692&y=-37.9899781428645&sizex=650&sizey=650&level=6&star=1&circle=
[4] http://maps.google.com.au/maps?f=q&hl=en&q=dandenong+station,+vic&sll=-25.335448,135.745076&sspn=56.255093,84.023438&layer=&ie=UTF8&z=16&om=1
[5] Vincent Adams Winter (1990). *VR and VicRail: 1962 - 1983*. p. 106. ISBN 0 9582069 3 3.
[6] Fiddian, Mark (1997). *Trains, Tracks, Travelers. A history of the Victorian Railways*. South Eastern Independent Newspapers. p. 154. ISBN 1 875475 12 5.
[7] "Upgrading Eltham to a Premium Station". *Newsrail* (Australian Railway Heritage Society (Victorian Division)): page 310. October 1997.
[8] Department of Infrastructure website (http://www.doi.vic.gov.au/DOI/Internet/transport.nsf/AllDocs/7D244319338C6A0ACA2572FA0005B349?OpenDocument)
[9] http://vicsig.net/index.php?page=infrastructure&name=Dandenong

The closed **General Motors station** is located towards Hallam station on the Pakenham line, and the proposed **Lynbrook station** is located towards Cranbourne, as is the freight only **Lyndhurst station**.

South_Gippsland

Farmland near Kilcunda

Regions of Gippsland
East West South Central Latrobe Valley

South Gippsland, a region of Gippsland in Victoria, Australia, is a well-watered region consisting of low, rolling hills descending to the coast in the south and the Latrobe Valley in the north. Low granite hills continue into Wilsons Promontory, the southernmost point of Victoria and mainland Australia. Rivers are generally very short and impossible to dam owing to the lack of potential storage sites, but groundwater of good quality is readily available. The major industries are forestry and dairy farming, and the principal towns include Cowes (on Phillip Island), Leongatha, Korumburra, Wonthaggi and Foster.

Wilsons Promontory National Park features eucalypt forests and rainforests as well as its famous beaches, and is one of the most popular holiday areas in Victoria. Linked to mainland South Gippsland via a bridge at San Remo, Phillip Island is also a major tourist destination, noted particularly for its surf beaches, nightly Penguin Parade and Grand Prix track. 12,000 years ago, South Gippsland formed part of a land bridge to Tasmania the remnants of which is the Furneaux Group of islands. A ferry operates from Welshpool to Lady Barron Island, part of the Furneaux Group.

Korumburra_railway_station

Korumburra	
Station Information	
Line	South Gippsland Railway
Distance from Flinders Street	116km
Number of Platforms	1
Number of Tracks	6
Station status	Heritage and **Tourist Station**
Station opened	2 June 1891
Station closed	24 July 1993

Korumburra is a heritage Listed railway station on the former South Gippsland line in South Gippsland, Victoria, Australia. The station has been part of the South Gippsland tourist railway since passenger operations on the line ceased beyond Cranbourne station in July 1993.

History

Between 1999 and 2008 there was constant speculation that the railway line beyond Cranbourne to Leongatha could re-open, as promised by the Victorian State Government, under a project named 'Bringing Trains Back to Victorians'.

However, in May 2008, a scoping study carried out on behalf of the State Government found the costs of returning services high, at $72 million dollars. Therefore, plans to reopen the line were stopped, and the Government has pledged to spend $14.2 million dollars on improved V/Line coach services in the South Gippsland region instead.

Further, there are plans in motion to turn the railway reservation into a Rail Trail between Cranbourne East and Nyora. [1]

Location

Korumburra Railway Station is located on Station Street, just off Bridge Street, Korumburra. Bridge street can be accessed from the South Gippsland Highway Bridge Street is named after the road overpass bridge that crosses the railway line, dividing the eastern and western halves of the town.

The station contains a well preserved red brick station, large goods yard, a fully working turntable and goods shed. The station building currently enjoys a heritage listing, and the station's 100th birthday was in January 2008.

Korumburra Railway Station features a Railway Museum, featuring railway artifacts and items from the entire South Gippsland Line. Various items have been donated from people throughout the district. An additional ongoing display is a photographic exhibition featuring images of railway infrastructure, vehicles and railway workers in action from Koo-Wee-Rup through to Foster and beyond - all prior to the 1994 closure of the line.

As well as the many freight trains that ran on the line superphosphate special trains were commonly stabled at the Korumburra Yards. Superphosphate trains would run around Winter months and transported many goods products, farming produce, fertilizer, oil, explosives, super, grain, cement and many more products.

Public transport services

Upon the closure of the railway stations for regular passenger traffic, the V/Line coach stop was located outside the Korumburra station in Station Street.

Changes to services and streamlining of coach services meant the change of V/Line coach stops along the route, with Korumburra's coach stop being moved from the Korumburra Station to the main street.

Station layout

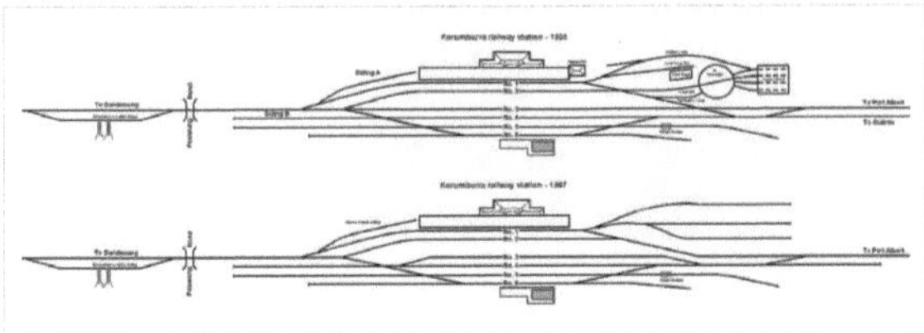

References

[1] Communication from Minister Lynne Kosky, Dated June 3, 2008

External links

- http://www.sgr.org.au/about.html

Korumburra,_Victoria

Korumburra Victoria	
 Commercial Street	
[[file:|270px|Korumburra is located in]]<div style="position: absolute; z-index: 2; top: %; left: %; height: 0; width: 0; margin: 0; padding: 0;"> Korumburra	
Population:	4,465(2006)[1]
Postcode:	3950
Coordinates:	38°26′S 145°49′E
Location:	• 120 km (75 mi) SE of Melbourne • 15 km (9 mi) W of Leongatha
LGA:	South Gippsland Shire
County:	Buln Buln
State District:	Gippsland South
Federal Division:	McMillan

Korumburra is a town in the Australian state of Victoria. It is located on the South Gippsland Highway, 120 kilometres (75 mi) kilometres south-east of Melbourne, in the South Gippsland Shire local government area.

Surrounded by rolling green hills, the town is 227 metres (745 ft) above the sea level of coastal Inverloch, about 28 km (17 mi) away. At the 2006 census, Korumburra had a population of 4465.

History

The name is believed to be derived from the Aboriginal word meaning 'blow-fly'.

The Post Office in the area opened on 1 September 1884, and moved to the township on the railway survey line on 1 November 1889, the existing office being renamed Glentress.[2] The railway arrived in 1891.

In the winter of 2005, Korumburra was blanketed in snow for the first time in almost twenty years. Local residents were seen skiing the tenth fairway at the Korumburra Golf Club.[3]

The town has also enjoyed a wave of migration of European migrants who have added to the towns growth and culture. Antonio Radovick "Father of Korumburra" was the most successful Croatian pioneer in Victoria who contributed to the start and growth of the town in the 1890s and 1900s. There is a street bearing the name Radovick near the middle hotel. One of the old Korumburra Shire Council shire presidents was Italian migrant Vincenzo Fava [1990] which shows the contribution of Italians in Korumburra.

The Town today

Korumburra is known as the "Heritage Centre of South Gippsland". It is the home of Coal Creek Community Park and Museum. This village depicts life in the area over the period from the 1870s to 1920s, as the town rapidly expanded following the discovery of a coal seam. The outdoor museum covers 30 acres (m^2) of bushland, including 60 exhibits. These include the Giant Earthworm, National Bank, Anzac exhibit, Mining exhibits, Dairy exhibit at the Boston Carriageworks and Railway Museum. The Park is free entry and open Thursdays to Mondays during termtime and seven days a week during school holidays. A tramway runs on weekends around the lower end of the park encompassing an old-time farm and bush oval. Many local organisations use the Park and environs and special events are held during the year.(www.coalcreekvillage.com.au)

Korumburra is also the base of the popular tourist railway South Gippsland Railway which operates a heritage railway service between the major centres of Leongatha and the market town of Nyora (formerly the junction to the Wonthaggi line). Heritage trains are operated each Sunday and on public holidays. The South Gippsland Tourist Railway have operated since 1994 when the line was closed in the 90's. The last V/Line passenger train ran to Korumburra on the Leongatha line on Saturday the 24th of July 1993. The line beyond Leongatha is being used as a rail trail for public use and also the former Wonthaggi line. Dandenong - Cranbourne is being used by the Melbourne Suburban train company, while the section beyond Cranbourne - Nyora is in an unusable state for trains to operate and is yet to have its fate decided yet.

Other town attractions include the Olympic pool (open November–March), two-court basketball stadium, art gallery, the South Gippsland tourist railway. The towns main industries include dairy and beef. The region is home to the world's largest earthworms. The town is also home to Burra foods which is a dairy company.

The town in conjunction with neighbouring township Bena has an Australian Rules football team competing in the Alberton Football League.[4]

The town has a soccer team competing in the South Gippsland Soccer League under the name Korumburra City.

Golfers play at the course of the Korumburra Golf Club on Warragul Road.[5]

The town is now being developed with large areas of former farm land being developed for new residential estates, which within the next ten years will expand the size and population of the township by 75 to 100 percent.

In DC Comics, Korumburra, spelled Kurrumburra, is the home town of The Flash's notorious nemesis, Captain Boomerang.

Earthquakes

On 6 March 2009 an earthquake registering 4.6 on the Richter scale was recorded 7 km (4 mi) west of Korumburra.[6] A second magnitude 4.5 tremor was recorded two weeks later on 18 March 2009; the epicentre was 5 km (3 mi) north of the town.[7] No damage was reported.[8]

As of 2 April 2009, fifteen earthquakes and aftershocks have been recorded around the town during 2009.[9]

On 5 July 2011 a magnitude 4.4 earthquake with the epicentre on Korumburra was felt over much of suburban Melbourne as well.[10]

See also

- Korumburra railway station

References

[1] Australian Bureau of Statistics (25 October 2007). "Korumburra (State Suburb)" (http://www.censusdata.abs.gov.au/ABSNavigation/prenav/LocationSearch?collection=Census&period=2006&areacode=SSC25915&producttype=QuickStats&breadcrumb=PL&action=401). *2006 Census QuickStats*. . Retrieved 2007-10-01.

[2] Premier Postal History. "Post Office List" (https://www.premierpostal.com/cgi-bin/wsProd.sh/Viewpocdwrapper.p?SortBy=VIC&country=). . Retrieved 2008-04-11

[3] Rachel, Kleinman; Shtargo, Sasha (11 August 2005). "Victoria's feeling for snow" (http://www.theage.com.au/news/national/victorias-feeling-for-snow/2005/08/10/1123353386875.html). The Age. . Retrieved 2007-08-04.

[4] Full Points Footy. "Korumburra-Bena" (http://www.fullpointsfooty.net/Korumburra-Bena.htm). . Retrieved 2008-07-25

[5] Golf Select. "Korumburra" (http://www.golfselect.com.au/armchair/courseView.aspx?course_id=1002). . Retrieved 2009-05-11

[6] "Recent Earthquake : N of Korumburra VIC" (http://www.ga.gov.au/bin/earthquake.pl?title=N+of+Korumburra+VIC&magnitude=4.6&depth=15&xy=145.839,-38.445&date=06,03,2009&time=09,55,36&bg1=eqrisk_lm&zoom=250&station=TOO). . Retrieved 2009-04-02.

[7] "Recent Earthquake : Korumburra, VIC" (http://www.ga.gov.au/bin/earthquake.pl?title=Korumburra,+VIC&magnitude=4.6&depth=18&xy=145.800,-38.390&date=18,03,2009&time=05,28,20&bg1=eqrisk_lm&zoom=250&station=TOO). 18 March 2009. . Retrieved 2009-04-02.

[8] The Age Melbourne hit by tremor (http://www.theage.com.au/national/melbourne-hit-by-tremor-20090318-91w6.html) March 18, 2009. Retrieved March 18, 2009.

[9] "Recent earthquakes measured by Geoscience Australia" (http://www.ga.gov.au/urban/listquakes.php?month=03&year=2009). . Retrieved 2009-04-02.

[10] "Earthquake report from Geoscience Australia" (http://www.ga.gov.au/earthquakes/getQuakeDetails.do?quakeId=3070931&orid=438695&sta=TOO). .

Victorian_Railways

Victorian Railways train restored to the livery used in the 1960's.

The **Victorian Railways** operated railways in the Australian state of Victoria from 1859 to 1983. The first railways in Victoria were private companies, but when these companies failed or defaulted, the Victorian Railways was established to take over their operations. Most of the lines operated by the Victorian Railways were of 1600 mm (5 ft 3 in) broad gauge, however the railways also operated up to five 762 mm (2 ft 6 in) narrow gauge lines between 1898 and 1962, and a 1435 mm (4 ft 8 $\frac{1}{2}$ in) standard gauge line between Albury and Melbourne from 1961.

History

Formation

A Department of Railways was created in 1856 with the first appointment of staff. British engineer, George Christian Darbyshire was made first Engineer-in-Chief in 1857, and steered all railway construction work until his replacement by Thomas Higginbotham in 1860. Because of political turmoil in the Victorian Government, Higginbotham was one of 137 officials removed from office on "Black Wednesday" on 8 January 1878 when the Government was denied supply. He, like a number of other senior officers, was not reappointed.[1] [2] Robert Watson then took over as Engineer-in-Chief. But in 1880 a new Ministry expressed a wish to redress the injustice by re-instating Higginbotham. At the sudden death of Higginbotham in 1880, William Elsdon took over for 2 years before his retirement in 1882, and Watson then return to his former position as Engineer-in-Chief, which he held up to the time of his death.

On 1 November 1883 assent was given to the Victorian Railways Commissioners Act 1883, 47 Vic., No.767, to construct, maintain and manage the state's railways. The staff of the Department of Railways came under the authority of the Railway Commissioners, which became commonly known as Victorian Railways. The elaborate headquarters at 67 Spencer Street were opened in 1893.[3]

Growth

Victorian Railways grew to service all parts of the state, even extending some lines into New South Wales under the 1922 Border Railways Act. In the late 19th century the railways became something of a political football with politicians demanding new lines to be built in places where traffic levels never justified it. In 1864 there was just 254 miles (409 km) of railway. The system expanded rapidly to reach 2,900 route miles (4,670 km) by 1891 and to its greatest extent of 4,755 route miles (7652 km) in 1939. The result was that by the beginning of the 20th century, no Victorian (apart from those in the mountain regions) were more than 25 miles (42 km) from a railway line. The period from the end of the 1930s saw a slow decline in route mileage as unprofitable branches were closed.

The *Spirit of Progress* headed by locomotive S301 *Sir Thomas Mitchell* near Kilmore East in 1938. (Photo: The late John L. Buckland)

Conversion of the Melbourne suburban system to electric operation commenced in 1919 and was completed by 1930, creating what was claimed at the time to be the world's largest electric suburban rail system. 1937 saw the introduction of the streamlined *Spirit of Progress* passenger train, with air conditioning and all steel carriage construction. Diesel power was introduced in 1951 with ten F-class diesel-electric shunting locomotives, followed by B-class mainline diesel-electric locomotives in 1952-3. A standard gauge line connecting to the New South Wales system was constructed in 1961 allowing through trains to operate between Melbourne and Sydney, Australia's two largest cities, for the first time. The last steam locomotive was withdrawn in 1972.

Head office at 67 Spencer Street.

Demise

In May 1973 the Railways (Amendment) Act 1972 passed the management of the Railways from the Victorian Railways Commissioners to a Victorian Railways Board. In 1974 the Victorian Railways was rebranded as VicRail, but the royal blue and gold livery used on rolling stock was retained until 1981.[4]

In 1983 VicRail was divided into two—the State Transport Authority taking responsibility for the provision of country rail and road, passenger and freight services, and the Metropolitan Transit Authority taking over suburban passenger operations.

The State Transport Authority traded under the V/Line name, while the Metropolitan Transit Authority used that name until the Public Transport Corporation ("The Met") was formed in 1989. Between 1996 and 1999 "V/Line" and "The Met" were privatised. "V/Line Passenger" was franchised to the National Express Group, who withdrew in 2002. The former "V/Line Freight" division was sold to Freight Victoria and is now owned by Pacific National, V/Line's former interstate rail freight infrastructure is controlled by the Australian Rail Track Corporation, and Metro Trains Melbourne now operates the suburban railway network.

Management

When first formed in 1857, the management of the Railways Department was initially vested in the President of the Board of Land and Works, this situation remaining until 1884.[5] With the passing of the Victorian Railways Commissioners Act 1883, a board of four commissioners was put in charge, responsible to the Minister of Railways (the Minister of Transport from 1935 onwards).[6]

Harold Winthrop Clapp

The Chairman of Commissioners of the Victorian Railways were:[7] [8]

- Richard Speight: 1883 to 1892
- Richard Hodge Francis: 1892 to 1894
- James Syder: 1894 to 1896
- John Mathieson: 1896 to 1901
- William Francis Joseph Fitzpatrick: 1901 to 1903
- Thomas James Tait: 1903 to 1910
- William Francis Joseph Fitzpatrick: 1910 to 1915
- Charles Ernest Norman: 1915 to 1920
- Harold Winthrop Clapp: 1920 to 1939
- Norman Charles Harris: 1940 to 1950
- Robert George Wishart: 1950 to 1955
- Edgar Henry Brownbill: 1956 to 1967
- George Frederick Brown: 1967 to 1973

After the Bland Report of 1972, in May 1973 the Railways (Amendment) Act 1972 passed the management of the Railways from the Victorian Railways Commissioners to a Victorian Railways Board. The board could have up to seven members, with six being initially appointed. This remained until 1983 when the board was discontinued under the Transport Act 1983.[8]

Fleet

The Victorian Railways operated a wide variety of locomotives and rolling stock to provided passenger and goods services. This included equipment acquired from the private companies that built the first railways in Victoria. The majority was the fleet was broad gauge, with a specialised fleet used on the narrow gauge lines. In later years gauge conversion was used to place stock from the main VR fleet onto standard gauge.

First generation diesel locomotive B60 beside one of the S class steam locomotives it replaced.

The first locomotives used in the state were small steam locomotives, the majority being imported from Britain, with later years seeing larger units being built locally.[9] Electric locomotives were acquired with the electrification of the suburban railways, with more powerful units acquired when the mainline to Traralgon was electrified.[9] Dieselisation occurred from 1951, but the B class of 1952 revolutionised main line operations.[9] Apart from the F class shunters, Clyde Engineering had a monopoly on Victorian diesel-electric locomotives,[10] as the Australian licensee of General Motors EMD engines and traction motors, fitting them into locally designed bodies.

Early passenger services were operated with 4 and 6 wheeled 'dogbox' passenger carriages, but larger bogie rollingstock started to appear from the turn of the century.[11] On the Melbourne suburban network electric multiple units were introduced speeding up services. Experiments were also made with various diesel and petrol railcars for use on smaller branch lines. By the late 1970s, country passenger services were run down, and older wooden rolling stock was now approaching their use by date. As a result, the *New Deal* saw modern steel carriages introduced from 1981.[11]

Early wagons were built on four wheeled under frames, but from 1871 bogie vehicles begun to appear.[12] The last four wheeled open wagons were built in 1958,[13] but were not scrapped in large numbers until the 1980s when new bogie wagons replaced them,[14] by 1987 the bogie wagon fleet numbered 5000.[14]

When the Victorian Railways (now known as VicRail) was divided into two in 1983, the Metropolitan Transit Authority received the suburban electric multiple unit fleet, while the State Transport Authority took responsibility for remainder for the provision of country passenger and freight services.

Operational branches

The Victorian Railways was divided up into a number of branches, each with a set of responsibilities. These branches were reorganised a number of times, in 1962 they were:[15]

- Secretaries: headed by the Secretary for Railways, dealt with policy, administration, transport regulation and legal matters.
- Rolling Stock: headed by the Chief Mechanical Engineer, was responsible for design, construction, operation, and maintenance of all locomotives and rolling stock.
- Way and Works: headed by the Chief Civil Engineer, it constructed and maintained all fixed infrastructure such as track, bridges, stations, signalling and safeworking.
- Traffic: headed by the Chief Traffic Manager, it operated all goods and passenger services both on rail and road.
- Electrical Engineering: headed by the Chief Electrical Engineer, it managed the suburban railway electrification system, as well as power supply to stations.
- Accountancy: headed by the Controller of Accounts, it records all payments, prepared budgets, conducted audits, and paid salaries and wages to employes.

- Commercial: headed by the Chief Commercial Manager, it set goods rates and passenger fares, solicited for new traffic to rail, and took action against by-laws offenders.
- Stores: headed by the Controller of Stores, it received all incoming stores and materials, and controlled the railway printing works.
- Refreshment Services: headed by the Superintendent of Refreshment Services, it controlled food and bookstore services at station, managed advertising at stations, as well as the railway bakery, butchery, poultry farm and laundry.

Visual identification

For most of the 20th century, the colours of royal blue and gold were the distinctive feature of the Victorian Railways. It was first introduced on the *Spirit of Progress* express train in 1937 along with the winged 'VR' logo, and was refined to the final form with the arrival of the B class diesel electric locomotives in 1952. The revised logo was inspired by that of the Erie Railroad in the United States.

Early version of the Victorian Railways 'wings' as used on the *Spirit of Progress.*

While the *Spirit of Progress* carriages wore the royal blue and gold striping, the remainder of the passenger fleet wore a more plain red livery. Additional carriages did not appear in the blue and gold until the 1954 Royal Tour by HM Queen Elizabeth II. Freight stock was painted in a slightly different red / brown with only identifying lettering painted in white on the side.

VicRail 'teacup' logo of 1981

In 1961 with the coming of the standard gauge line into Victoria, the Victorian Railways held a competition to find a "symbol, sign or slogan" to be carried on new freight vehicles for the line. The winner was an 18 year old art student from Bentleigh, with the logo being a stylised VR with arrowheads either end. By the 1970s most bogie vehicles wore the logo, until May 1983 pending the launch of V/Line.[16]

In 1974 the Victorian Railways was rebranded as VicRail, with a new logo unveiled on April 12, 1976,[17] but the royal blue and gold image was retained until 1981,[4] when the orange and silver 'teacup' scheme was launched on locomotives, Comeng trains, and passenger carriages. This was the last livery, with V/Line launched in August 1983 with a "stylised capital lettered logo with the V and the L split by a deep slashing stroke".[16]

Named trains

The Victorian Railways operated a number of named passenger trains, including the:[18]

- *Geelong Flyer*
- *Great Northern Limited*
- *The Overland*
- *The Boat Train*
- *Spirit of Progress*
- *The Gippslander*
- *Intercapital Daylight*
- *Mildura Sunlight*
- *Southern Aurora*

- *The Northener*
- *Vinelander*
- *The West Coaster*
- *Fruit Flyer*

The railways also operated a number of specialist trains that were used to bring services to rural and isolated populations. These included:

- *Reso Train*
- *Better Farming Train*
- *Train of Knowledge*

Other functions

From 1888 the Victorian Railways began to take on a role in tourism, operating the Victorian Government Tourist Bureaux until it was taken over by the state government in 1959.[19] In 1911 the Victorian Railways Commissioners assumed responsibility for the State Coal Mine at Wonthaggi from the Mines Department.[20]

The railways also operated the Newport Power Stations A and B, railway refreshment services, the Mount Buffalo Chalet guesthouse, road motor services for passengers, and motor transport services for goods. The railways also operated two tram routes in Melbourne—the "Electric Street Railways".[21] [22]

References

- Public Record Office Victoria - Victorian Railways [23]

[1] Brian Harper relating conversation with descendants of John Darbyshire

[2] Sands & McDougall's Melbourne and Suburban Directory, 1865, 1870, 1885

[3] Geoff Peterson (February 1993). "67 Spencer Street". *Newsrail* (Australian Railway Historical Society (Victorian Division)): pages 44–45.

[4] Railmac Publications (1992). *Australian Fleetbooks: V/Line locomotives*. Kitchner Press. p. 5. ISBN 0 949817 76 9.

[5] Leo J. Harrigan (1962). *Victorian Railways to '62*. Public Relations and Betterment Board. p. 273.

[6] Leo J. Harrigan (1962). *Victorian Railways to '62*. Public Relations and Betterment Board. p. 274.

[7] Leo J. Harrigan (1962). *Victorian Railways to '62*. Public Relations and Betterment Board. p. 275.

[8] Vincent Adams Winter (1990). *VR and VicRail: 1962 - 1983*. pp. 6–9. ISBN 0 9582069 3 3.

[9] Railmac Publications (1992). *Australian Fleetbooks: V/Line locomotives*. Kitchner Press. pp. 2–3. ISBN 0 949817 76 7.

[10] "VR - V/Line - VLP/ FA Locomotives" (http://locopage.railpage.org.au/vline.html). *Locopage*. . Retrieved 2008-02-05.

[11] "Victorian Railways - Loco hauled Passenger Carriages" (http://www.pjv101.net/cd/pages/type_car.htm). *Peter J. Vincent's Website*. . Retrieved 2008-02-08.

[12] Norm Bray and Peter J Vincent (2006). *Bogie Freight Wagons of Victoria*. Brief History Books. ISBN 0 9775056 0 X.

[13] Mark Bau. "Four wheeled open wagons of the Victorian Railways". *Notes from the Victorian Model Railway Society Prototype Modellers Meet 2007*.

[14] "V/Line Freight Rollingstock Fleet - 1 July 1987". *Newsrail* (Australian Railway Historical Society (Victorian Division)) **15** (10): page 303. July 1987.

[15] Leo J. Harrigan (1962). *Victorian Railways to '62*. Public Relations and Betterment Board. p. 162.

[16] Norm Bray and Peter J Vincent (2006). *Bogie Freight Wagons of Victoria*. Brief History Books. p. 14. ISBN 0-9775056-0-X.

[17] Vincent Adams Winter (1990). *VR and VicRail: 1962 - 1983*. p. 205. ISBN 0 9582069 3 3.

[18] Leo J. Harrigan (1962). *Victorian Railways to '62*. Public Relations and Betterment Board. pp. 256–268.

[19] Leo J. Harrigan (1962). *Victorian Railways to '62*. Public Relations and Betterment Board. p. 168.

[20] Leo J. Harrigan (1962). *Victorian Railways to '62*. Public Relations and Betterment Board. p. 165.

[21] "V.R. Tramway "Reminisences."" (http://www.tramway.org.au/reflections.php?p=vr_tramway_reminisences). *TMSV Running Journal*. . Retrieved 2010-11-28

[22] "The Sandringham Tramway" (http://www.tramway.org.au/reflections.php?p=the_sandringham_tramway). *TMSV Running Journal*. . Retrieved 2011-03-06

[23] http://access.prov.vic.gov.au/public/component/daPublicBaseContainer?component=daViewAgency&breadcrumbPath=Home/Access%20the%20Collection/Browse%20The%20Collection/Agency%20Details&entityId=2876

Victorian_Heritage_Inventory

The **Victorian Heritage Inventory**, commonly known as the Heritage Inventory, is a list of all known historical archaeology sites in Victoria, Australia. It is maintained by **Heritage Victoria** [1] , the Victorian State Government's principal cultural (non-Indigenous) heritage agency

The Heritage Inventory, which lists over 5000 sites, can be searched online through the **Victorian Heritage Database.** [2]

References

[1] "Heritage Victoria" (http://www.heritage.vic.gov.au/). . Retrieved 2010-09-13.
[2] "Victorian Heritage Database" (http://vhd.heritage.vic.gov.au/). . Retrieved 2010-09-13.

List_of_closed_railway_stations_in_Victoria

This is a list of **former railway stations and railway lines in Victoria**, Australia. Many of these stations and lines have been abandoned or demolished. For closed railway stations in suburban Melbourne, see List of closed railway stations in Melbourne.

This transport-related list is incomplete; you can help by expanding it [1].

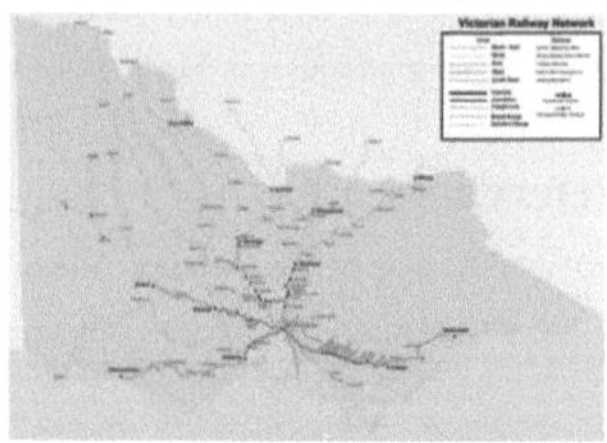

Overview map.

Passenger lines proposed for reopening

The Mildura line, whilst closed to passenger rail services at present has seen State Government proposals for their reintroduction. The line is currently open to freight; while an upgrade to the line was announced in 2006, this is limited to improving the current 40 km/h (25 mph) maximum speed on many sections,[2] and is not relevant to any hopes that a passenger train service might recommence.

In December 2008 as part of the Victorian Transport Plan the state government announced that Maryborough station would see 14 V/Line passenger rail services extended to the station every week, commencing 2010 at a cost of $50 million.[3]

There was originally a planned upgrade to a fourth line: the Leongatha line in South Gippsland. This plan was dropped in May 2008 after progress had stalled for some years.[4] [5] A tourist line, the South Gippsland Tourist Railway, continues to operate along a section of the line. The Leongatha line still has a chance in the future of reopening. The Liberals said if they win the next Victorian election, they will reopen the line if the transport crisis gets worse. The Labour Government said in May last year they might leave the track from Cranbourne - Nyora if needed in the future. But it has a chance of becoming a rail trail. Cranbourne - Nyora line is going to be left as a disused railway line. Leongatha and Cobram have been talked about to reopen in the future.

Mildura line

Once served by the *Vinelander* service, the line branches from the Serviceton line at Ballarat.

- **Ballarat**
- Creswick
- Clunes
- Maryborough
- Dunolly
- St Arnaud
- Donald
- Birchip
- Ouyen
- Red Cliffs
- Irymple
- **Mildura**

Closed lines

The following lines are either closed to regular passenger services, freight only, part of tourist and heritage railways, or closed and removed.[6]

- **Northern**:
- Melbourne - Bendigo:
  - Lancefield railway line
  - Daylesford railway line (Daylesford Spa Country Railway)
  - Castlemaine-Maryborough railway line
  - Maldon railway line (Victorian Goldfields Railway)
  - Redesdale railway line
  - Heathcote railway line
- West from Bendigo:
  - Robinvale railway line (freight only)
  - Wedderburn railway line
  - Eaglehawk-Inglewood railway line
  - Kulwin railway line (freight only)
  - Stony Crossing railway line
  - Yungera railway line
  - Kerang-Koondrook Tramway
- East from Bendigo:
  - Cohuna railway line
  - Toolamba-Echuca railway line (freight only)
  - Deniliquin railway line (freight only)
  - Balranald railway line (freight only to Moulamein in New South Wales)
- **North Eastern**:
- Melbourne - Wodonga:
  - Heathcote railway line
  - Mansfield railway line
  - Tatong railway line
  - Oaklands railway line, Victoria (freight only)

- Whitfield railway line
- Bright railway line
- Yackandandah railway line
- Peechelba railway line
- Wahgunyah railway line
- Cudgewa railway line (freight only to Bandiana)

- Goulburn Valley:
  - Rushworth railway line
  - Toolamba-Echuca railway line (freight only)
  - Katamatite railway line (freight only to Dookie)
  - Picola railway line
  - Cobram railway line
  - Tocumwal railway line (freight only beyond Shepparton)
- **Midland**:
  - Geelong-Ballarat railway line (freight only)
  - Gheringhap-Maroona railway line (freight and *The Overland* only)
  - Mildura railway line (freight only)
- Ballarat region:
  - Waubra railway line
  - Ballarat Cattle Yards railway line
  - Buninyong railway line
  - Skipton railway line
  - Newtown-Colac railway line
- Ballarat - Serviceton:
  - Avoca railway line
  - Navarre railway line
  - Carpolac railway line
  - Bolangum railway line
  - Patchewollock railway line (freight only to Hopetoun)
  - Yaapeet railway line (freight only)
  - Yanac railway line
- Ballarat - Mildura:
  - Avoca railway line (freight only)
  - Castlemaine-Maryborough railway line (freight only to Moolort from Maryborough)
  - Pinnaroo railway line (freight only)
  - Millewa South railway line
  - Morkalla railway line
  - Ballarat to Daylesford railway line
- Ararat - Portland:
  - Portland railway line (freight only)
  - Grampians railway line
  - Balmoral railway line
  - Koroit-Hamilton railway line
  - Coleraine railway line
  - Casterton railway line

  - Mount Gambier railway line (isolated by rail gauge)
- **South Western**:
  - Geelong-Ballarat railway line
  - Fyansford railway line
  - Cunningham Pier railway line
  - Queenscliff railway line (Bellarine Peninsula Railway)
  - Geelong Racecourse railway line
  - Wensleydale railway line
  - Forrest railway line
  - Newtown-Colac railway line
  - Crowes railway line
  - Alvie railway line
  - Timboon railway line
  - Mortlake railway line
  - Koroit-Hamilton railway line
  - Port Fairy railway line (open as far as Warrnambool)
- **Eastern**:
- Melbourne - Orbost:
  - Orbost railway line (open as far as Bairnsdale)
  - Noojee railway line
  - Walhalla railway line (Walhalla Goldfields Railway)
  - Thorpdale railway line
  - Yallourn railway line
  - Mirboo North railway line
  - Maryvale railway line
  - Maffra railway line
  - Briagolong railway line
  - Sale Wharf railway line
  - Bairnsdale Wharf railway line
- South Gippsland:
  - Strzelecki railway line
  - Wonthaggi railway line
  - Outtrim railway line
  - Barry Beach railway line
  - Welshpool Jetty railway line
  - Woodside railway line
  - South Gippsland railway line (South Gippsland Railway)

Closed stations

On operational lines

South West

Name	Line	Location	Opened	Closed	Current Status
Moriac	Warrnambool line	Moriac	?	Late 1970s	Demolished
Pirron Yallock	Warrnambool line	Pirron Yallock	?	Late 1970s	Platform demolished, building stands
Panmure	Warrnambool line	Panmure	?	Late 1970s	Demolished
Allansford	Warrnambool line	Allansford	?	1981	Demolished

West

Name	Line	Location	Opened	Closed	Current Status
Parwan	Ballarat line	Parwan-Exford Road, Parwan	?	1980s	Demolished
Rowsley	Ballarat line	Rowsley	?		Demolished
Llandello	Ballarat line	Ballan	?		Demolished
Gordon	Ballarat line	Gordon	?	Late 1970s	Intact, leased out
Millbrook	Ballarat line	Millbrook	1885[7]	1898	Demolished
Wallace	Ballarat line	Wallace	1885[8]	1898	
Bungaree	Ballarat line	Bungaree	?	1981	Platform intact
Dunnstown	Ballarat line	Dunnstown	1885[9]	1898	Platform demolished, goods shed remains
Warrenheip	Ballarat line	Warrenheip	?	Late 1970s	Demolished
Burrumbeet	Ararat line	Burrumbeet	?	?	Platform intact
Trawalla	Ararat line	Trawalla	?	Late 1970s	Intact
Buangor	Ararat line	Buangor	?	Late 1970s	Intact, but in disrepair

North

Name	Line	Location	Opened	Closed	Current Status
Rupertswood	Bendigo line	Within Sunbury	1870s	December 2004	Intact, one platform demolished.
Carlsruhe	Bendigo line, former junction of Daylesford line	?, Carlsruhe	?	Destaffed 1978, closed 1982	Intact, building leased out.
Taradale	Bendigo line	Station Street, Taradale	?	1976	Intact, leased out.
Elphinstone	Bendigo line	?, Elphinstone	?	Late 1970s	Intact, building leased out.
Harcourt	Bendigo line	?, Harcourt	?	Late 1970s	Intact, leased out
Golden Square	Bendigo line	?, Golden Square	?	Late 1970s	Intact, building leased out.
Raywood	Swan Hill line	Raywood	?	Late 1970s	?

Mitiamo	Swan Hill line	Mitiamo	?	Late 1970s	?
Macorna	Swan Hill line	Macorna	?	Late 1970s	?
Lake Boga	Swan Hill line	Lake Boga	?	Late 1970s	Demolished

North East

Name	Line	Location	Opened	Closed	Current Status
Mangalore	Shepparton line	Mangalore	?	Early 1980s	Demolished
Tabilk	Shepparton line	Tabilk	?	Late 1980s	Demolished
Wahring	Shepparton line	Wahring	?	Late 1970s	Demolished
Arcadia	Shepparton line	Arcadia	?	Late 1970s	Demolished
Toolamba	Shepparton line	Toolamba	?	Late 1970s	Demolished
Longwood	Albury line	Longwood	November 20, 1872	October 4, 1981	Demolished
Baddaginnie	Albury line	Baddaginnie	June 19, 1882	July 5, 1978	Demolished, goods platform still visible
Glenrowan	Albury line	Glenrowan	November 2, 1874	October 4, 1981	Demolished, rebuilt 2001
Barnawartha	Albury line	Barnawartha	November 21, 1873	Late 1970s	Intact

East

Name	Line	Location	Opened	Closed	Current Status
Fernbank	Bairnsdale line	Fernbank	?	Late 1970s	Demolished
Lindenow	Bairnsdale line	Lindenow	?	Late 1970s	Demolished
Hillside	Bairnsdale line	?	?	Late 1970s	Demolished

On closed lines

Name	Line	Location	Opened	Closed	Current Status
Kensington	Queenscliff line	? Leopold	1880	1881	Demolished
Queenscliff Junction	Queenscliff line	? South Geelong	1879	1901	Demolished
Queenscliff line	? Breakwater	1891	?	Demolished	
Scarborough	Queenscliff line	?	1889	1914	Demolished
Cheethams Siding	Queenscliff line	Whittington	1909	1959	Demolished
Moolap	Queenscliff line	Moolap Station Road, Moolap	1881	1961	Demolished
Leopold	Queenscliff line	Leopold	1881 as Kensington Flat, renamed in 1886	1961	Demolished
Curlewis	Queenscliff line	Curlewis	1914	1961	Demolished
Mannerim	Queenscliff line	Mannerim	1883, renamed from Marcus Hill station in 1890	1961	Demolished

Marcus	Queenscliff line	? Marcus	1883 as Ocean Grove station, renamed in 1896 as Marcus Hill and renamed again in 1904	1961	Demolished
Tylden	Daylesford line	?, Tylden	?	1978	Demolished
Fern Hill	Daylesford line	?	?	1978	Demolished
Woodleigh	Wonthaggi line	?	1910	late 1977	Intact, Demolished
Kernot	Wonthaggi line	?	1910	late 1977	Intact, Demolished
Almurta	Wonthaggi line	?	1910	late 1977	Intact, Demolished
Glen Forbes	Wonthaggi line	?	1910	late 1977	Intact, Demolished
Woolamai	Wonthaggi line	?	1910	late 1977	Intact, Demolished
Anderson	Wonthaggi line	?	1910	late 1977	Intact, Demolished
Kilcunda	Wonthaggi line	?	1910	late 1977	Intact, Demolished
Dalyston	Wonthaggi line	?	1910	late 1977	Intact, Demolished
Wonthaggi	Wonthaggi line	?	1910	late 1977 - 78	Intact, Used (Station Has been transformed into Local Museum)
Clyde	South Gippsland line	Railway Rd, Clyde	1892	1994	Intact, Demolished
Tooradin	South Gippsland line	Tooradin Station Road, Tooradin	1892	1994	Intact, Demolished
Koo Wee Rup	South Gippsland line	Station Road, Koo Wee Rup	1892	1994	Services to be restored in the future.
Monomeith	South Gippsland line	Monomeith Road, Monomeith	1892	1994	Intact, Demolished
Lang Lang	South Gippsland line	Rosebery Road, Lang Lang	1892	1994	Services to be restored in the future.
Ruby	South Gippsland line	Ruby Road, Ruby	1892	1994	Intact, Demolished
Koonwarra	South Gippsland line	Johnsons Road Koonwarra	1892	1994	Intact, Demolished
Meeniyan	South Gippsland line	Scholers Road, Meeniyan	1892	1994	Intact, Demolished
Buffalo	South Gippsland line	Neals Road, Buffalo	1892	1994	Intact, Demolished
Fish Creek	South Gippsland line	Falls Road, Fish Creek	1892	1994	Intact, Demolished
Foster	South Gippsland line	Station Road, Foster	1892	1994	Intact, Demolished
Toora	South Gippsland line	Victoria Street, Toora	1892	1994	Intact, Demolished
Barry Beach Junction	South Gippsland line	Main Access Road (Off Barry Road) Barry Beach Junction	1892	1994	Marine Terminal Operational, Transported by Boat (No Rail), Rails Removed (Paved Over)
Welshpool	South Gippsland line	Railway Avenue, Welshpool	1892	1994	Intact, Demolished
Yeoburn	Kerang-Koondrook Tramway line	Yeoburn	1889	1981	?
Hinksons	Kerang-Koondrook Tramway line	?	1889	1981	?

Teal Point	Kerang-Koondrook Tramway line	Teal Point	1889	1981	?
Gannawarra	Kerang-Koondrook Tramway line	Gannawarra	1889	1981	?
Koondrook	Kerang-Koondrook Tramway line	Koondrook	1889	1981	?
Broomfield	Ballarat to Daylesford line	Broomfield	?	?	?
Allendale	Ballarat to Dalyesford line	Allendale	?	?	?
Kingston	Ballarat to Dalyesford line	Kingston	?	?	?
Newlyn	Ballarat to Dalyesford line	Newlyn	?	?	?
Rocklyn	Ballarat to Dalyesford line	Rocklyn	?	?	?
Wombat	Ballarat to Dalyesford line	Wombat	?	?	?
Leonard	Ballarat to Dalyesford line	Leonard	?	?	?
Sailors Falls	Ballarat to Dalyesford line	Sailors Falls	?	?	?
Woodburn	Ballarat to Dalyesford line	Woodburn	?	?	?

See also

- List of regional railway stations in Victoria
- List of closed Melbourne railway stations

References

[1] http://en.wikipedia.org/wiki/List_of_closed_railway_stations_in_victoria

[2] Department of Transport. "Mildura Rail Freight Upgrade" (http://www.transport.vic.gov.au/DOI/Internet/Freight.nsf/AllDocs/122B8C7141DD6801CA2573290018B59A?OpenDocument). www.transport.vic.gov.au. . Retrieved 2008-07-14.

[3] "Maryborough Rail Extension Projects" (http://www4.transport.vic.gov.au/vtp/projects/maryborough.html). *Victorian Transport Plan.* www4.transport.vic.gov.au. . Retrieved 2008-12-08.

[4] TransNet Logistics (May 1, 2007). "Leongatha Rail Line: Reintroduction of Passenger Train Services (Review Summary & Transport Options Assessment Report)" (http://www.doi.vic.gov.au/DOI/DOIElect.nsf/$UNIDS+for+Web+Display/754A14B7BD86361FCA25743D0005919B/$FILE/Leongatha_Rail_Line_Review.pdf). www.doi.vic.gov.au. . Retrieved 2008-05-04.

[5] "GIPPSLAND PUBLIC TRANSPORT LINKS BOOSTED IN BUDGET" (http://www.doi.vic.gov.au/DOI/DOIElect.nsf/$UNIDS+for+Web+Display/493DA3882B071AD3CA25743D000CD078/$FILE/MediaRelease-SouthGippsland-020508.pdf). *Media release from the Minister for Public Transport.* www.doi.vic.gov.au. May 2, 2008. . Retrieved 2008-05-04.

[6] "Victorian Railway Maps 1860 - 2000" (http://www.vrhistory.com/VRMaps/index.htm). www.vrhistory.com. . Retrieved 2008-06-09.

[7] [http://www.vrhistory.com/Locations/A057-Gordon-Wallace.pdf Gordon-Wallace Victorian Railways History

[8] [http://www.vrhistory.com/Locations/A057-Gordon-Wallace.pdf Gordon-Wallace Victorian Railways History

[9] [http://www.vrhistory.com/Locations/A057-Gordon-Wallace.pdf Gordon-Wallace Victorian Railways History

External links

- Vicsig.net - statewide line guides (http://www.vicsig.net/index.php?page=infrastructure&section=lineguide)
- Timeline featuring opening and closure dates of lines (http://www.victorianrailways.net/vr history/history.html)
- Victorian Railway Maps 1860 - 2000 (http://www.vrhistory.com/VRMaps/index.htm)

Jumbunna_railway_station

Jumbunna	
Station information	
Line	Outtrim
Distance from Southern Cross	?
Number of platforms	?
Number of tracks	?
Station status	**Closed**
Station opened	1894
Station closed	1 October 1953

Jumbunna was a railway station on the Outtrim line in South Gippsland, Victoria, Australia. The station was opened in the 1890s, and was one of only three stations on the Outtrim line which was eventually closed in two stages between 1951 and 1953, when Victorian Railways was closing many other small branch lines.

The original station, now converted into a 3 bedroom house, and miniature railway was put up for sale in late 2006.[1]

See Also

- Jumbunna, Victoria

References

[1] http://www.alexscott.com.au/public/PropertyProductDisplay.asp?PID=25185&RURL=/postings/10331/25185.html&AID=&RID=&WAID=75&WRID=&UT=6/11/2006%202%3A22%3A00%20PM&RT=6/11/2006%202%3A22%3A00%20PM

Transport_in_Australia

There are many varying forms of **transport in Australia**.

Roads

The Eastern Suburbs Line passing over the Eastern Distributor motorway in Sydney

Australia has the second highest level of car ownership in the world. It has three to four times more road per capita than Europe and seven to nine times more than Asia. Australia also has the third highest per capita rate of fuel consumption in the world. Perth, Adelaide and Brisbane are rated among the most car-dependent cities in the world, with Sydney and Melbourne close behind.[1] Furthermore, the distance travelled by car (or similar vehicle) in Australia is among the highest in the world, being exceeded by USA and Canada.[2]

There are 3 different categories of Australian roads:

- Federal Highways
- State Highways
- Local Roads

The road network comprises a total of 913,000 km broken down into:[3]

- *Paved:* 353,331 km (including 3,132 km of expressways)
- *Unpaved:* 559,669 km (1996 estimate)

The majority of road tunnels in Australia have been constructed since the 1990s to relieve traffic congestion in metropolitan areas, or to cross significant watercourses.

Further information: List of tunnels in Australia

Public transport in Australia

Rising petrol prices and increasing traffic congestion are thought to be factors contributing to renewed growth in use of urban public transport.[4]

Intra-city public transport networks

Epping Station on Sydney's Cityrail network

The table below lists major cities in Australia with currently operating multi-modal intra-city (as opposed to inter-city or regional) public transportation networks.

The only Australian capital cities without such networks are Canberra and Darwin.

Trams in Australia historically serviced many Australian towns and several cities formerly operated tram networks, however the majority of these were shut down before the 1970s. Melbourne is an exception here however, and today boasts the largest tram network of any city in the world. Major regional cities where trams formerly facilitated multi-modal public transport networks Launceston, Geelong, Ballarat, Bendigo and Rockhampton.

Most major cities have at minimum bus services and these cities have been excluded services as have any with tourist or heritage transport (such as the private monorail at Sea World or the tourist Victor Harbor Horse Drawn Tram).

City	Overview	Integrated Network Name	Buses and taxicabs	Urban rail[5]	Light rail[6]	Monorail	Watercraft[7]
Adelaide	Public transport in Adelaide	Adelaide Metro	Overview	Overview	Overview		
Brisbane	Public transport in Brisbane	Translink	Overview	Overview			Overview
Darwin	Public transport in Darwin		✓				limited access
Gold Coast	Public transport on the Gold Coast	Translink	Overview	Overview	Overview	Overview	
Hobart	Public transport in Hobart	Metro	✓				✓
Melbourne	Public transport in Melbourne	Metlink	✓	✓	✓		limited access
Newcastle	Bus routes in Newcastle, New South Wales	Transport for NSW	✓	limited access			
Perth	Public transport in Perth	Transperth	✓	✓	✓		✓
Sydney	Public transport in Sydney	Transport for NSW	✓	✓	✓	✓	✓

Intercity rail transport

The railway network is large, comprising a total of 33,819 km (2,540 km electrified) of track: 3,719 km broad gauge, 15,422 km standard gauge, 14,506 km narrow gauge and 172 km dual gauge. Rail transport started in the various colonies at different dates. Privately owned railways started the first lines, and struggled to succeed on a remote, huge, and sparsely populated continent, and government railways dominated. Although the various colonies had been advised by London to choose a common gauge, the colonies ended up with different gauges.

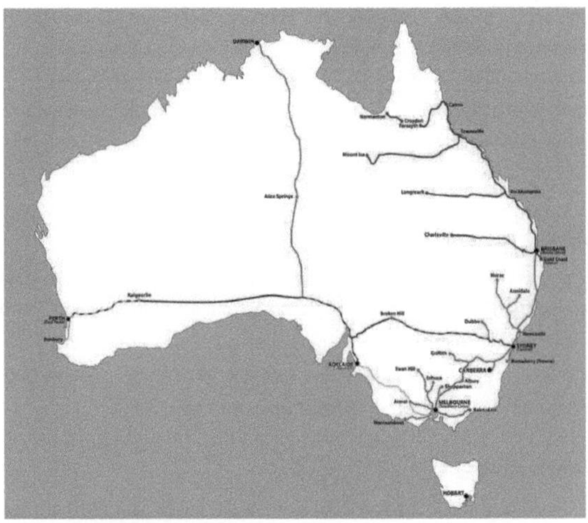

Map of passenger railway services in Australia State Government owned rail services: Queensland RailQR Citytrain and Traveltrain services CountryLink and CityRail services V/Line services Transwa servicesGreat Southern Railway (Australia)Great Southern Railway lines: Indian Pacific The Overland The Ghan

Inter-state rail services

The Great Southern Railway, owned by Serco Asia Pacific, operates three trains: the Indian Pacific (Sydney-Adelaide-Perth), The Ghan

(Adelaide-Alice Springs-Darwin), and The Overland (Melbourne-Adelaide) [8]. NSW owned CountryLink services link Brisbane, Canberra and Melbourne via Sydney. Since the extension of the Ghan from Alice Springs to Darwin was completed in 2004, all mainland Australian capital cities are linked by standard gauge rail, for the first time.

Looking along the Trans-Australian Railway

Intra-state and city rail services

There are various state and city rail services operated by a combination of government and private entities, the most prominent of these include V/Line (regional trains and buses in Victoria); Metro Trains Melbourne which operates the Melbourne rail network; RailCorp operating all passenger rail services in New South Wales including (CityRail and CountryLink);Queensland Rail (QR) operating Traveltrain and the Citytrain network, South-East Queensland's commuter railway network under the TransLink scheme, and Transwa operating train and bus services in Western Australia.

The Indian Pacific in Perth

Metro

Major cities in Australia do not have full-fledged rapid transit systems. Melbourne, Sydney, Brisbane and Perth's systems are all partially underground. Melbourne has plans for a new train service branded as a metro, but as it will interact with the suburban system and won't be grade-separated from Footscray out, it will fall short of the criteria of a metro. Plans for a "Euro-style" metro in Sydney have been shelved in favor of additional underground lines on the suburban network.

Mining railways

Four heavy-duty mining railways carry iron ore to ports in the northwest of Western Australia. These railways carry no other traffic, and are isolated by deserts from all other railways. The lines are standard gauge and are built to the heaviest US standards.

In 2006, a fifth iron ore railway is proposed by the Fortescue Metals Group, while a sixth common carrier railway is proposed to serve the port of Oakajee just north of Geraldton.

Cane railways

In Queensland about 15 sugar mills have narrow gauge (2 ft (610 mm) gauge) cane tramways that deliver sugar cane to the mills.

Pipelines

There are several pipeline systems including:

- *Crude oil:* 2,500 km
- *Petroleum products:* 500 km
- *Natural gas:* 5,600 km
- *Water*

- Perth to Kalgoorlie - Goldfields Water Supply Scheme
- Morgan on the Murray River to Adelaide, Whyalla, Port Lincoln
- Waranga Western Channel, near Colbinabbin to Bendigo and Ballarat - Goldfields Superpipe[9]

Projects under construction or planned:

Victoria

- Goulburn River to Sugarloaf Reservoir, Melbourne (North-South Pipeline, alternatively called the Sugarloaf Pipeline) - construction commenced 2008, expected completion in 2011.[10]
- Wimmera-Mallee Pipeline - construction commenced on supply systems 3 & 4 in May 2008, with other systems still under construction. Expected completion of all systems in 2010.[11]
- Melbourne to Geelong Pipeline - construction will commence in 2010, expected completion in 2011.[12]
- Rocklands Reservoir to Grampian Headworks pipeline (Hamilton - Grampians Pipeline) - construction commenced December 2008, expected completion in 2010.[13]

Waterways

Australia's inland waterways are not a significant means commercial transport. In the 19th century, paddle steamers were used on the Murray-Darling Basin to transport produce such as wool and wheat but the water levels are highly unreliable, making the river impassable for large parts of the year. The steamers proved unable to compete with rail, and later, road transport. Traffic now on inland waterways is therefore largely restricted to private recreational craft.

Ports and harbours

Mainland

General

- Adelaide, Brisbane, Cairns, Darwin, Fremantle, Geelong, Gladstone, Port Lincoln, Mackay, Melbourne, Newcastle, Portland, Sydney, Townsville, Wollongong

Ferries in Sydney

Iron ore

- Dampier
- Port Hedland
- Geraldton
- Oakajee - proposed 2006
- Esperance
- Port Lincoln - possible 2007

Tasmania

- Devonport
- Launceston
- Hobart

Merchant marine vessels

As of 2006, the Australian fleet consists of 53 ships of 1,000 gross register tons or over. The use of foreign registered ships to carry Australian cargoes between Australian ports is permitted under a permit scheme, with either Single Voyage Permit (SVP) or a Continuous Voyage Permit (CVP) being issued to ships.[14] Between 1996 and 2002 the number of permits issued has increased by about 350 per cent.[15]

Container crane and ship at the Port of Melbourne

Over recent years the number of Australian registered and flagged ships has greatly declined, from 75 ships in 1996 to less than 40 in 2007, by 2009 the number is now approaching 30. Marine unions blame the decline on the shipping policy of the Howard Government which permitted foreign ships to carry coastal traffic.[16] There have also been cases where locally operated ships have Australian flag from the vessel, registering it overseas under a flag of convenience, then hiring foreign crews who earn up to about half the monthly rate of Australian sailors.[15] Such moves were supported by the Howard Government but opposed by maritime unions and the Australian Council of Trade Unions.[17] The registration of the ship overseas also meant the earnings of the ships are not subject to Australian corporate taxation laws.[16]

Statistics for the Shipping Industry of Australia	
Total: 53 ships (1,000 gross register tons (GRT) or over)	
Totalling: 1,361,000 GRT/1,532,874 metric tons deadweight (DWT)	
Cargo ships	
Bulk ships	17
Cargo ship	4
Container ships	1
Roll-on/Roll-off ships	5
Tankers	
Liquefied gas tanker ships	4
Chemical tanker ships	3
Petroleum tanker ships	6
Passenger ships	
General passenger ships	6
Combined passenger/cargo	7
Source: *This article contains material from the CIA World Factbook which, as a US government publication, is in the public domain.*	

Airports

Further information: List of the busiest airports in Australia

Major Australian Airlines:

- Qantas
  - Jetstar (subsidiary)
- Virgin Australia
- Tiger Airways Australia

Melbourne (Tullamarine) Airport

Sydney (Kingsford-Smith) Airport

There are many airports around Australia paved or unpaved. A 2004 estimate put the number of airports at 448. The busiest airports in Australia are:

1 Sydney Airport Sydney, New South Wales SYD

2 Melbourne Airport Melbourne, Victoria MEL

3 Brisbane Airport Brisbane, Queensland BNE

4 Perth Airport Perth, Western Australia PER

5 Adelaide Airport Adelaide, South Australia ADL

6 Gold Coast Airport Gold Coast, Queensland OOL

7 Cairns Airport Cairns, Queensland CNS

8 Canberra International Airport Canberra, Australian Capital Territory CBR

9 Hobart International Airport Hobart, Tasmania HBA

10 Darwin International Airport, Northern Territory DRW

Airports with paved runways

Total: 305

- *Over 3,047 m (10,000 ft):* 10
- *2,438 to 3,047 m (8,000 to 10,000 ft):* 12
- *1,524 to 2,437 m (5,000 to 8,000 ft):* 131
- *914 to 1,523 m (3,000 to 5,000 ft):* 139
- *Under 914 m (3,000 ft):* 13 (2004 estimate)

Airports with unpaved runways

*Total:*143

- *1,524 to 2,437 m (5,000 to 8,000 ft):* 17
- *914 to 1,523 m (3,000 to 5,000 ft):* 112
- *Under 914 m (3,000 ft):* 14 (2004 estimate)

Note:

sourced from CIA World Fact Book https://www.cia.gov/library/publications/the-world-factbook/index.html"

Environmental impact

The environmental impact of transport in Australia is considerable. Australia subsidizes fossil fuel energy, keeping prices artificially low and raising greenhouse gas emissions due to the increased use of fossil fuels as a result of the subsidies. The Australian Energy Regulator and state agencies such as the New South Wales' Independent Pricing and Regulatory Tribunal set and regulate electricity prices, thereby lowering production and consumer cost.

See also

- Inland Railway
- Lonie Report
- Economy of Australia
- Telecommunications in Australia

References

[1] http://www.eoc.csiro.au/lb/lbbook/urban/yb2.htm
[2] "Transport in Australia" (http://www.iraptranstats.net/aus). iRAP. . Retrieved 2009-02-17.
[3] CIA world fact book
[4] http://www.abs.gov.au/ausstats/abs@.NSF/94713ad445ff1425ca25682000192af2/d81efef6e2252cf4ca256f7200833049!OpenDocument
[5] Includes electrified networks only
[6] includes modern tram networks
[7] includes public ferry and Water taxi services
[8] http://www.gsr.com.au/trains.htm
[9] http://www.ourwater.vic.gov.au/programs/water-grid/goldfields
[10] http://www.ourwater.vic.gov.au/programs/water-grid/sugarloaf
[11] http://www.ourwater.vic.gov.au/programs/water-grid/wimmera-mallee
[12] http://www.ourwater.vic.gov.au/programs/water-grid/geelong-melbourne
[13] http://www.ourwater.vic.gov.au/programs/water-grid/hamilton-grampians
[14] Australian Shipowners Association. "Industry Policy" (http://www.asa.com.au/industrypolicy.asp). www.asa.com.au. . Retrieved 2009-11-08.
[15] Paul Robinson (26 March 2002). "Maritime unions slam use of 'cheap' foreign labour" (http://www.theage.com.au/articles/2002/03/25/1017004765455.html). *The Age*. www.theage.com.au. . Retrieved 2009-11-08.
[16] Martin Byrne (22 October 2009). "A new tanker ship for Australia" (http://www.aimpe.asn.au/files/caltex_tanker_letter_to_fed_gov.pdf). *Letter from the Australian Institute of Marine and Power Engineers to the Federal Minister*. www.aimpe.asn.au. . Retrieved 2009-11-08.
[17] Liz Porter (14 July 2002). "Shipping out, and definitely not shaping up" (http://www.theage.com.au/articles/2002/07/13/1026185123874.html). *The Age*. www.theage.com.au. . Retrieved 2009-11-08.

Sources

- *This article incorporates public domain material from the CIA World Factbook* document "2000 edition" (http://www.umsl.edu/services/govdocs/wofact2000/index.html).

External links

- Australia Transport Statistics (http://www.iraptranstats.net/aus)
- Railway maps (http://www.railpage.org.au/railmaps/)
- Driving time calculator for NSW (http://www.rta.nsw.gov.au/cgi-bin/index.cgi?action=fatiguecalc.form)

Outtrim_North_railway_station

Outtrim North	
Station information	
Line	Outtrim
Distance from Southern Cross	?
Number of platforms	?
Number of tracks	?
Station status	**Closed**
Station opened	1896
Station closed	4 September 1951

Outtrim North was a railway station on the Outtrim line in South Gippsland, Victoria, Australia. The station was opened in 1896, and was one of only three stations on the Outtrim line which was eventually closed in the two stages between 1951 and 1953, when Victorian Railways was closing many other small branch lines.

Bendigo_railway_line

Bendigo railway line, Victoria	
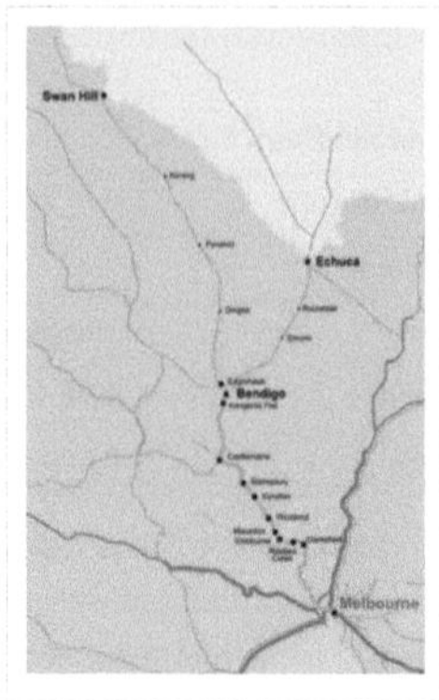	
Line details	
Length	164.24 km (102 mi)
Stations	11
Tracks	Four tracks to Footscray, double track to Kyneton, single line with crossing loops beyond
Rolling stock	VLocity
Rail transport in Victoria	
List of stations **Closed railways**	

The **Bendigo railway line** is a regional railway in Victoria, Australia. The electrified part of it within metropolitan Melbourne is the Sydenham line. It has 11 stations. The line was upgraded as part of the Regional Fast Rail project between 2005 and 2006.

History

Originally called the *Melbourne and Murray River Railway* it was constructed by the Victorian Railways Department under the supervision of Engineer in Chief George Christian Darbyshire and completed under Thomas Higginbotham. The line originating from Spencer Street Station reached Sunbury in 1859. It had reached Woodend and Kyneton by 1861 and Castlemaine in 1862. It opened in Bendigo on 20 October 1862.[1] The line includes substantial wrought iron and masonry viaducts at Sunbury, Malmsbury and Taradale, as well as two tunnels at Elphinstone and Big Hill.

Branch lines

The Lancefield branch line was opened from Clarkefield (north of Sunbury) to Lancefield in 1881, and extended to Kilmore in 1892 to connect with the Heathcote railway line. This line was completely closed by 1956.

The Daylesford branch line was opened from Carlsruhe (between Woodend and Kyneton) to Daylesford in 1880. This line was closed in 1978. Part of this line, between Daylesford and Bullarto is now operated by the Daylesford Spa Country Railway as a tourist railway. This line was connected with a line from Ballarat in 1887.

A branch line was built between Redesdale Junction (north of Kyneton) and Redesdale by 1900. It was closed in the 1950s.

The Maldon branch line was opened from Castlemaine to Maldon in 1884 and Shelbourne in 1891, although the railway had originally been planned to run to Laanecoorie. The line from Maldon to Shelbourne was closed in 1969 due to bush fire damage. The branch line closed in 1976.[2] The Victorian Goldfields Railway has restored the line and operates trains between Castlemaine and Maldon.

A branch line was built from Bendigo to Heathcote in 1888 and connected to a line from Kilmore in 1890. The Bendigo – Heathcote line closed in 1958. The Heathcote Junction to Heathcote branch was closed in November 1968.

Echuca line

The Bendigo line was extended to Elmore and Echuca in 1864 and across the Murray River to connect with the private Deniliquin and Moama Railway Company from Moama to Barnes and Deniliquin in 1876. This company was taken over by Victorian Railways in 1923.

A branch line was built from Elmore to Cohuna in 1910 and it was closed in the 1980s.

A branch line was built from Barnes to Moulamein and Balranald in 1926. The Moulamein – Balranald section was closed in the 1980s.

In 1996 the passenger service to Echuca was reinstated for the first time since 1983, when a twice-weekly service from Bendigo was started. Since 2007, there is one train to/from Melbourne on weekdays and two on weekends, with the train speed between Bendigo and Echuca limited to 80 km/h.

Swan Hill line

The Swan Hill line was extended north from Eaglehawk (just north of Bendigo on the line to Inglewood) in 1882, reaching Swan Hill in 1890. It remains in use today.

Robinvale line

The Robinvale line was opened from Bendigo to Inglewood in 1876, Korong Vale in 1882, Boort in 1883, Quambatook in 1894, Ultima in 1900, Chillingolah in 1909, Manangatang in 1914, Annuello in 1921 and Robinvale in 1924. This line currently only handles grain trains.

Victorian Railways commenced construction of a railway to Koorakee and Lette in New South Wales in 1924, but this railway was never completed. The Murray River bridge between Robinvale and Euston was instead converted to a road bridge. A new road bridge is currently being built to replace it and the old bridge will be demolished on its completion. A short branch line was built from Wedderburn Junction (south of Korong Vale) to Wedderburn in the 1880s and it was closed in the 1980s.

There is currently no passenger service on this line.

Kulwin line

The Kulwin line was opened from Korong Vale to Wycheproof in 1883, Sea Lake in 1895, Nandaly in 1914, Mittyack in 1919 and Kulwin in 1919.

This line currently only handles grain trains. Until late 2006, rural rail network lessee Pacific National had mothballed the Mittyack to Kulwin section but this has been recently re-opened to traffic despite the poor grain harvest. There has not been passenger service on this line since before 1984.

Line guide

Regular V/Line passenger services operate along the Bendigo line. Some services continue beyond Bendigo as the Swan Hill and Echuca lines. Suburban Metro Trains Melbourne services operate to Watergardens as the Sydenham line, the current limit of electrification on the line.

The portal of the *Big Hill* railway tunnel, 390 metres long

There is double track as far as Kyneton station, with a single track line continuing beyond. The line beyond Kyneton was previously double track prior to the Regional Fast Rail project in 2005.[3]

During peak hour some services terminate at Sunbury or Kyneton stations. Passengers on the inner section of the line are permitted to use Metcard tickets to access the services, with this section marked as the Sunbury line on suburban network maps.

Bold stations are termini, *italic* stations are staffed at least part time (this has been confirmed).

References

[1] http://www.arhsvic.org.au/content/railway-history-victoria-1839-1900
[2] "Chart of Events" (http://www.vgr.com.au/timechart/timechart1988.htm). Victorian Goldfields Railway (http://www.vgr.com.au). . Retrieved 2006-06-09.
[3] Media Release, Minister for Public Transport, Victoria, Monday November 17, 2003 (http://www.legislation.vic.gov.au/domino/Web_Notes/newmedia.nsf/b0222c68d27626e2ca256c8c001a3d2d/cc4932f92392b214ca256de10078b4f8!OpenDocument)

A mix of new and old signalling at Castlemaine station

External links

- http://www.vline.com.au (http://www.vline.com.au)
- Official map (http://www.vline.com.au/maps/maps/north.html)
- Statistics and detailed schematic map (http://www.vicsig.net/index.php?page=infrastructure&section=lineguide&line=Bendigo) at the vicsig (http://www.vicsig.net/) enthusiast website

Outtrim_railway_station

Outtrim	
Station information	
Line	Outtrim
Station status	**Closed**
Station opened	1896
Station closed	4 September 1951

Outtrim was the terminus station on the Outtrim railway line. It closed during the 1950s with the other 2 stations on the line. The first sod of soil turned in the construction of the Outtrim railway line was on 6 March 1885. The line was completed by 1896.

See Also

- Outtrim, Victoria

Mildura_railway_line

Mildura railway line, Victoria	
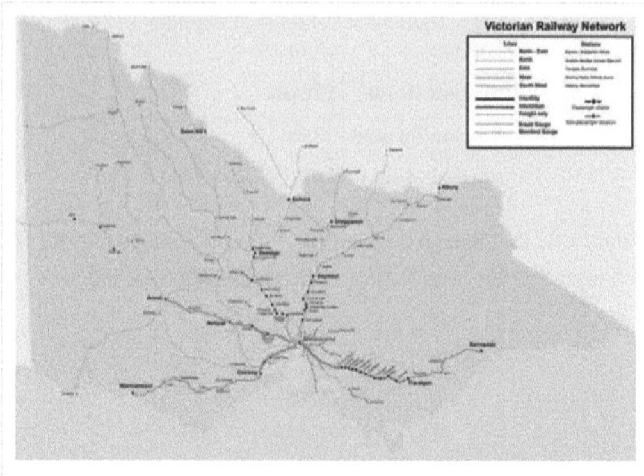	
Line details	
Commenced	1874
Completed	1903
Closed	September 12, 1993 (for passengers)
Reopened	July 2010 (to Maryborough)
Fate	Freight only; Passenger to Maryborough
Length	604 km (375 mi)
Stations	9
Tracks	Single track
Used by	Pacific National/El Zorro/V/Line (To Maryborough)
Connections	Robinvale/Kulwin line at Dunolly
Rail transport in Victoria	
List of stations **Closed railways**	

The **Mildura line** is a railway line in Victoria, Australia. It runs from Ballarat in an approximate north-norwesterly direction to Mildura. Initial sections of the line opened from Ballarat in 1874 and the line reached Mildura in 1903. It has nine main stations. Passenger services ceased on the line in 1993, but freight services continue to be operated on the line by Pacific National. An announcement was made in 2006 for passenger services to be re-introduced to Mildura after a 13 year absence, under the Linking Victoria program. Passenger services were reintroduced from Ballart to Maryborough in July 2010. [1] [2]

History

The Mildura line was opened from Ballarat to Creswick, Clunes, Maryborough and Dunolly in 1874 and 1875, and extended to St Arnaud in 1878, Donald in 1882, Birchip in 1893, Woomelang in 1899, Ouyen, Red Cliffs and Mildura in 1903, Merbein in 1910 and Yelta in 1925.

The expansive Maryborough station

A line was opened from Ballarat to Waubra in the 1880s. It was closed in the 1960s.

A branch line was built from North Creswick to Daylesford in 1887, connecting with the line from Carlsruhe. This line was closed in 1953.

A line was opened from Maryborough to connect with the Melbourne - Bendigo line at Castlemaine in 1874. This line is not currently used by regular traffic.

Another branch line was opened from Maryborough to Avoca in 1876 and Ben Nevis and Ararat in 1890. This line was closed in 1959, reopened in 1966, standardised in 1996 and closed again in 2005. A branch was opened from Ben Nevis to Navarre in 1914 and it was closed in 1954.

A line was opened from Dunolly to Inglewood in 1888, connecting with the Bendigo - Boort line, completed in 1883.

A line was opened from Ouyen to Murrayville in 1912 and the South Australian border in 1915, connecting with the line to Pinnaroo and Tailem Bend. The line west of Pinnaroo was standardised in 1998 creating a break-of-gauge. Nevertheless, it is proposed to standardise the line from Geelong to Mildura and the line between Ouyen and Pinnaroo. In 2006, state and federal money was announced to upgrade the line to Mildura with gauge convertible sleepers. However, this upgrade will still leave the passenger service in limbo.

A branch line was completed from Red Cliffs to Meringur in the 1920s and extended to Morkalla in the 1930s. This line was closed in the 1980s. A 2 km section of the line near Red Cliffs has been converted to a 610 mm-gauge steam railway, operated occasionally by the **Red Cliffs Historical Steam Railway** (see Victorian Tourism [3]).

In 1923, a series of sidings were placed between the Mildura railway station and the wharves on the Murray River. These included a zig-zag section to enable trains to travel between the different elevations. These sidings were removed in 1973.[4]

A passenger service was provided by V/Line with three weekly overnight services from Melbourne (known as the *Vinelander*), including sleeping cars, until 1993 when it was discontinued by the Kennett Government.[5] The former Bracks/Brumby state Labor Governments promised to re-open this line to passenger traffic as part of its pledges for the 1999, 2002, 2006 and 2010 elections.

In December 2007 work began on upgrading the line between Gheringhap (on the Geelong - Ballarat line and Mildura, at a cost of $73 million and involving one in every two sleepers being replaced, and lifting maximum train speeds to 80 km/h.[6] Current speeds are restricted to 50 km/h on one-third of the track and are as low as 30 km/h in some sections.[7] In April 2008 it was announced that the Mildura - Yelta section of the line would also be upgraded, as part of the Victorian core grain network in a $23.7 million package with 6 other lines.[8]

As part of the upgrades for the return of Maryborough passenger services, the crossing loops at Sulky, Tourello and Talbot stations have been removed, resulting in only one train at a time being able to use the 60 kilometres of line between Ballarat and Maryborough.[9] The loop at Sulky was removed on February 12, the loop at Talbot on February 24.[10]

In October 2010, the Victorian government released a report into public transport improvement options for the north-west of Victoria, outlining nine proposals for public transport projects. Options discussed include the return of passenger trains on the existing line, or via an extension the extension of the Swan Hill railway line to connect to the

Mildura line at Ouyen.[11] [12]

Services

The main freight traffic on the line is export grain and containerised wine, grapes, citrus, dried fruit and juice, totalling around 1.5 million tonnes per year. Containers are dispatched to Melbourne from a terminal at Merbein (a short distance north of Mildura) operated by Wakefield Transport,[13] which handles approximately 13,000 export containers a year, as well as 500 import containers.[14] Cement is also despatched to Mildura from Waurn Ponds (near Geelong), oil and less than container load (LCL) freight were carried until 2007.[15] A freight terminal also exists at Donald, and grain services run as required.

In December 2008 as part of the Victorian Transport Plan the state government announced that V/Line rail passenger services would be extended from Ballarat to Maryborough station at a cost of $50 million,[16] commencing July 25, 2010.[17] Initially Creswick was to be the only intermediate station,[18] [19] until June 2010 when the government backflipped and announced $7 million in funding to also reopen Clunes.[20]

See also

- Tourist and Heritage Railways Act

References

[1] "V/Line Annual Report 2009-2010" (http://www.vline.com.au/pdf/publications/annualreports/annualreport09-10.pdf). V/Line Pty Ltd. September 2010. p. 7..

[2] "Transport Projects - Maryborough Rail Services" (http://www.transport.vic.gov.au/web23/Home.nsf/AllDocs/2e119eebf81209d2ca257625001ae39c?OpenDocument). *Department of Transport website*. Department of Transport. . Retrieved 2 July 2011.

[3] http://www.visitvictoria.com/displayObject.cfm/objectid.0003C002-D4F8-1A7B-835580C476A90000/vvt.vhtml

[4] Bruce Maclean (February 1978). "Mildura's Riverfront Railways". *Australian Railway Historical Society Bulletin* (Australian Railway Historical Society): pages 33–38.

[5] "The Mildura Story The Battle of the Routes" (http://www.historyvictoria.org.au/pdf/Mildura.pdf). Royal Historical Society of Victoria. . Retrieved 2008-08-30.

[6] "WORKS BEGIN ON MILDURA RAIL FREIGHT UPGRADE" (http://www.minister.infrastructure.gov.au/aa/releases/2007/December/AA001_2007.htm). www.minister.infrastructure.gov.au. . Retrieved 2008-04-27.

[7] "Roads, ports and freight - Mildura Rail Freight Upgrade - Frequently asked questions" (http://www.doi.vic.gov.au/Doi/Internet/Freight.nsf/AllDocs/CEFAA4D3E21F849DCA2573290018A76E?OpenDocument). www.doi.vic.gov.au. . Retrieved 2008-04-27.

[8] "$43m to upgrade rail freight lines" (http://business.theage.com.au/43m-to-upgrade-rail-freight-lines/20080421-27nl.html). business.theage.com.au. . Retrieved 2008-04-27.

[9] V/Line Weekly Notice: "NORTH BALLARAT - MARYBOROUGH. TOURELLO LOOP ABOLISHED LEVEL CROSSING UPGRADES"

[10] "Signalling and Infrastructure". *Railway Digest* (Australian Railway Historical Society (NSW Division)): page 56. May 2010.

[11] [http://www.premier.vic.gov.au/component/content/article/12274.html Victorian Government press release - "NORTH WEST PUBLIC TRANSPORT REVIEW REPORT RELEASED", October 13th 2010

[12] North West Public Transport Review (http://transport.vic.gov.au/DOI/Internet/transport.nsf/AllDocs/AB4C036AA57FC455CA2571FA008328EE?OpenDocument)

[13] "Roads, ports and freight - Intermodal terminals" (http://www.doi.vic.gov.au/DOI/Internet/Freight.nsf/AllDocs/0B68D60810012C21CA256E05000488F9?OpenDocument). www.doi.vic.gov.au. . Retrieved 2008-08-01.

[14] "Melbourne–Mildura Corridor Strategy" (http://www.infrastructure.gov.au/transport/publications/files/Melbourne_Mildura_Corridor_Strategy.pdf). *Building our National Transport Future*. 74.125.153.132. June 2007. . Retrieved 2009-11-03.

[15] "Wongm's Rail Gallery - Mildura freight" (http://wongm.railgeelong.com/mildura-freight). wongm.railgeelong.com. . Retrieved 2009-11-03.

[16] "Maryborough Rail Extension Projects" (http://www4.transport.vic.gov.au/vtp/projects/maryborough.html). *Victorian Transport Plan*. www4.transport.vic.gov.au. . Retrieved 2008-12-08.

[17] "Maryborough Rail Services" (http://www.transport.vic.gov.au/web23/Home.nsf/AllDocs/2e119eebf81209d2ca257625001ae39c?OpenDocument). *Victorian Transport Plan*. www.transport.vic.gov.au. . Retrieved 2010-07-05.

[18] KIM STEPHENS (October 7, 2009). "Call to explain Clunes' omission from rail plan" (http://www.thecourier.com.au/news/local/news/general/call-to-explain-clunes-omission-from-rail-plan/1643848.aspx). *The Courier*. www.thecourier.com.au. . Retrieved 2010-06-15.

[19] EMMA BROWN (October 11, 2009). "No plans for Clunes train" (http://www.thecourier.com.au/news/local/news/general/no-plans-for-clunes-train/1646383.aspx). *The Courier*. www.thecourier.com.au. . Retrieved 2010-06-15.

[20] "READY FOR TOMORROW: $19 MILLION TO RETURN PASSENGER RAIL TO CLUNES AND IMPROVE NORTH-WEST FREIGHT RAIL LINKS" (http://www.premier.vic.gov.au/newsroom/10767.html). *Media Release: Premier of Victoria*. www.premier.vic.gov.au. June 2010. . Retrieved 2010-06-15.

External links

- Statistics and detailed schematic map (http://www.vicsig.net/index.php?page=infrastructure&section=lineguide&line=Yelta) at the vicsig (http://www.vicsig.net/) enthusiast website

North_East_railway_line

North East railway line, Victoria	
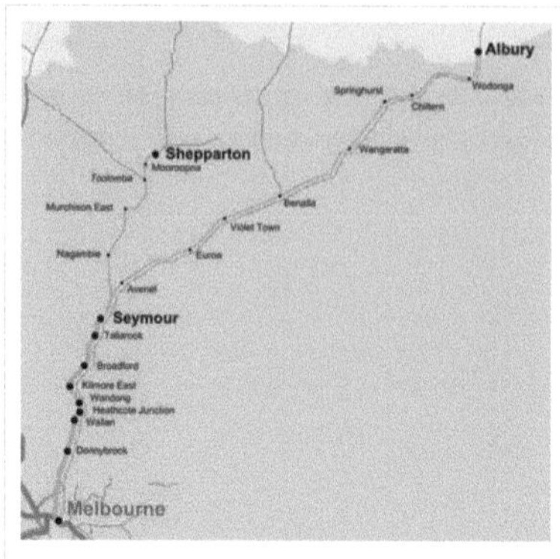	
Line details	
Commenced	1860
Completed	1883
Tracks	Two 1600 mm (5 ft 3 in) (broad gauge) tracks to Seymour with a single BG track continuing to Shepparton; as well as a single standard gauge track from Melbourne to Seymour with 2 tracks from Seymour to Albury
Used by	V/Line passenger services and intrastate freight on the broad gauge, CountryLink XPT and interstate freight on the standard gauge
Connections	Seymour and Shepparton lines
Rail transport in Victoria	
List of stations **Closed railways**	

The **North East railway** is a railway serving the north-east of Victoria, Australia.[1] Running from the state capital Melbourne to the New South Wales railways at Albury, it also serves the regional cities of Seymour, Benalla, Wangaratta and Wodonga. Between Seymour and Albury, the route consists of two parallel standard gauge tracks managed by the Australian Rail Track Corporation, one of which was recently a broad gauge line operated by V/Line.[2] The line also includes a broad gauge line from Melbourne to Shepparton. At Albury the line joins up with the New South Wales' Main South Line from Sydney.

Services

On the broad gauge line Metro Trains Melbourne operates suburban passenger services along the inner section of the line as the Craigieburn Line, while V/Line services operate as the Interurban Seymour Line and Shepparton Lines. Freight services also use the line, operated by Pacific National.

The standard gauge line sees use by the CountryLink XPT, as well as being a major interstate freight link used by Pacific National, QRNational and Australian Railroad Group. Later in 2011 it is expected that V/Line will recommence services on the Intercity Albury-Wodonga with SG-converted rolling stock.

Both lines are owned by VicTrack, the broad gauge line managed by V/Line, and the standard gauge line by the Australian Rail Track Corporation.[3]

History

The Melbourne and Essendon Railway Company opened the first section of the Albury-Wodonga line from North Melbourne to Essendon in 1860.[4] Following its takeover by the Victorian Government in 1867, the line was extended by 1872[5] to School House Lane on the south side of the Goulburn River near Seymour, and later that year to Seymour and then to Longwood. Benalla, Wangaratta, Springhurst and Wodonga were reached in 1873,[4] connecting with the New South Wales Government Railways at Albury at a break-of-gauge in 1883. the design engineer was Robert Watson.

The steam-hauled *Spirit of Progress* near Kilmore East in 1937

Construction of a standard gauge track parallel with the broad gauge track from Albury to Melbourne was commenced in 1959, completing the Sydney-Melbourne railway.[6] The first freight train operated on the line on 3 January 1962,[7] followed by the first passenger train on 16 April the same year.[6]

B class diesel leading the *Spirit of Progress* in the 1950s

The line was once the home of a number of prestige passenger services between the state capitals of Melbourne to Sydney, including the *Spirit of Progress*, *Southern Aurora*, and *Intercapital Daylight*; but due to declining patronage were replaced by the CountryLink XPT from the 1990s.

Maintaining two parallel single track railways has drawn criticisms over the years, including inefficiencies in maintaining track, operating trains, and duplicated train control centres.[2] By 2001 the State Government announced the gauge conversion of the broad gauge line to standard,[8] but action was stifled due largely to complex leasing arrangements. Speed restrictions had been applied to the broad gauge line due to track deterioration.[9]

The *Spirit of Progress* in the 1970s

In May 2008 it was again announced that the line would be upgraded, with the conversion of 200 kilometres (120 mi) of broad gauge track to standard gauge between Seymour and Albury, construction of a 5 km (3.1 mi) bypass around Wodonga, and rail upgrades between Melbourne and Seymour carried out, including new passing loops.[10]

New passenger platforms will also be built on the standard gauge line, and V/Line locomotives and carriages will be gauge converted to operate on the line. Costing A$501.3 million, the Victorian Government is contributing A$171.3

million, the Australian Government is investing A$45 million in the Wodonga Rail Bypass, and the Australian Rail Track Corporation will contribute A$285 million and will take-on responsibility for the ongoing operation of the new north-east standard gauge rail line under a 45-year lease agreement with Victoria.[11] The project was due for completion by 2010, with passenger services disrupted for up to 12 months.[12]

On 8 November 2008, broad gauge passenger rail services on the line ceased after the running of the evening V/Line service from Melbourne to Albury and a special train, operated by the Seymour Railway Heritage Centre, forming the final passenger service to run from Albury to Melbourne.[13] [14] In December 2008 standardisation works commenced, being contracted by ARTC to the Southern Improvement Alliance. The first train to run along the new Wodonga Rail Bypass operated in March 2010.[15]

In early August 2010 CountryLink decided to terminate all Sydney-Melbourne XPT services at Albury for an indefinite length of time, due to defects in the newly resleepered track. "Mud holes" in the track resulted in speed restrictions on more than 200 kilometres (about 66 per cent) of the line between the Albury and Melbourne, adding an extra 1.5 hours to the travelling time.[16] Train drivers have blamed the ARTC's $285 million concrete sleeper project for the track issues, stating that the incorrect insertion of 300,000 new concrete sleepers is to blame. They have also repeatedly reported freight trains breaking couplings, due to the rough track.[17] CountryLink services did not resume until mid September 2010. V/Line passenger services are unlikely to return before the end of the year.[18]

Branch lines south of Seymour

A branch line was opened from Heathcote Junction (near Kilmore) to Kilmore in 1888 and extended to Tooborac in 1890, connecting to a line from Bendigo and Heathcote opened a little earlier. The Heathcote Junction – Heathcote line closed in 1968. A branch line was built from Kilmore to Lancefield in 1892, but it was closed in 1904.

The Mansfield line was opened from Tallarook to Yea in 1883, Molesworth in 1889 Cathkin and Merton in 1890 and Mansfield in 1891. This line is now closed. A branch line was built from Cathkin to Koriella in 1890 and Alexandra in 1909. This line closed in 1978.

Branch lines north of Seymour

The Shepparton line was opened from Mangalore to Toolamba and Shepparton in 1880.

A branch line was opened from Benalla to St James in 1883, Yarrawonga in 1886 and Oaklands in 1938, creating a break-of-gauge until the New South Wales Government Railways line was closed south of Boree Creek. Another short branch line was built from Benalla to Tatong in 1914, but closed in 1947.

The narrow-gauge Whitfield branch line was opened from Wangaratta to Whitfield in 1899, closing in 1953.

Two branch lines were opened from Bowser (north of Wangaratta) to Everton in 1875 and extended to Beechworth in 1876 and Yackandandah in 1891. This line closed in 1954. The Bowser – Everton line was extended to Myrtleford in 1883 and Bright in 1890. This line is now closed. A short branch line to Peechelba East also branched from Bowser

A short branch line was opened from Springhurst via Rutherglen to Wahgunyah in 1879. Services were suspended in 1995.

A branch line was opened from Wodonga to Tallangatta between 1889 and 1891, Shelley in 1916, Beetoomba in 1919 and Cudgewa in 1921.[19] A direct connection from Albury was added near Wodonga, creating a turning triangle to enable the *Sydney Limited* and its successor *Spirit of Progress* with their observation cars to be turned as a complete trainset. The line was closed beyond Bandiana in 1981,[6] and the connection to Wodonga later removed, with only standard gauge traffic continuing to use the line via Albury.[20]

Line guide

Bold stations are termini, *italic* stations are staffed at least part time. Broad gauge, 1600 mm (5 ft 3 in), is in red, and standard gauge, 1435 mm (4 ft 8 $^1/_2$ in), is in blue.

Junction of the North East and Shepparton line, site of the former Mangalore station

Standard gauge Pacific National operated container freight service near Seymour

V/Line Albury-Wodonga line service on the line

Steamrail "Farewell Broad Gauge" special, Benalla, 2008

References

[1] Turton, Keith W. (1973). *The North East Railway. (A Lineside Guide) Melbourne to Wodonga.* Australian Railway Historical Society (Victorian Division). ISBN 0858490137.

[2] "Derailment of Freight Train 1SP2N and the Subsequent Collision of Passenger Train 8318" (http://www.atsb.gov.au/publications/investigation_reports/2003/RAIR/rair2003002.aspx). *Australian Transport Safety Bureau: Investigation Reports.* 1 November 2006. . Retrieved 2008-03-03.

[3] "Background - Organisation of Australia's Railways" (http://www.infrastructure.gov.au/rail/trains/background/index.aspx). *Department of Transport and Regional Services.* . Retrieved 2008-03-03.

[4] Sid Brown (March 1990). "Tracks Across the State". *Newsrail* (Australian Railway Historical Society (Victorian Division)) **18** (3): pages 71–76.

[5] ***The Centenary of the Opening of the Railway to Seymour*** Turton, Keith W. Australian Railway Historical Society Bulletin, April;May 1972 pp73-95;101-109

[6] "VR timeline" (http://www.victorianrailways.net/vr history/history.html) (HTML). http://www.victorianrailways.net/. Mark Bau. . Retrieved 2008-03-03.

[7] Australian Railway Historical Society Bulletin, March 1962 pp35-39

[8] Victorian Auditor General (August 2006). "Rail Gauge Standardisation Project" (http://www.audit.vic.gov.au/reports_mp_psa/psa1603.html) (HTML). *Victorian Auditor General's Office.* . Retrieved 2008-02-08.

[9] V/Line: *V/Line Voice* (http://www.vline.com.au/media/newsletter/customer_newsletter.html) Issue 32, February 2008

[10] "$500m rail link upgrade for Victoria" (http://news.ninemsn.com.au/article.aspx?id=571907). news.ninemsn.com.au. . Retrieved 2008-05-30.

[11] "Premier of Victoria, Australia - FEDERAL-STATE CO-OPERATION DELIVERS MAJOR RAIL PROJECT" (http://www.premier.vic.gov.au/newsroom/federal-state-co-operation-delivers-major-rail-project.html). www.premier.vic.gov.au. . Retrieved 2008-05-30.

[12] "Full steam ahead - ABC News (Australian Broadcasting Corporation)" (http://www.abc.net.au/news/stories/2008/05/30/2260001.htm). www.abc.net.au. . Retrieved 2008-05-30.

[13] Victoria MacDonald (2008-11-08). "Historic day as a railway era comes to an end". *The Border Mail* (Albury-Wodonga: Fairfax Media)

[14] "Seymour Railway Heritage Centre - Tours 2008 - The Last Broad Gauge Passenger Train from Albury" (http://www.srhc.com.au/tours.php?action=display&id=39). Seymour Railway Heritage Centre. . Retrieved 2008-11-09.

[15] VICTORIA MACDONALD (16 March 2010). "A bypass milestone as freight train ambles in" (http://www.bordermail.com.au/news/local/news/general/a-bypass-milestone-as-freight-train-ambles-in/1777168.aspx). *The Border Mail.* www.bordermail.com.au. . Retrieved 2010-03-16.

[16] BRAD WORRALL (10 August 2010). "XPT passengers forced onto buses" (http://www.bordermail.com.au/news/local/news/general/xpt-passengers-forced-onto-buses/1909171.aspx). *The Border Mail.* www.bordermail.com.au. . Retrieved 2010-09-14.

[17] BRAD WORRALL (11 August 2010). "V/line mud 'chaos'" (http://www.bordermail.com.au/news/local/news/general/vline-mud-chaos/1909791.aspx?storypage=0). *The Border Mail.* www.bordermail.com.au. . Retrieved 2010-09-14.

[18] BRAD WORRALL (13 September 2010). "Trains back on track" (http://wodonga.iprime.com.au/index.php/news/prime-news/trains-back-on-track-video). *Prime Wodonga*. wodonga.iprime.com.au.. Retrieved 2010-09-14.
[19] Australian Railway Historical Society Bulletin, June 1960 pp91-93
[20] "Cudgewa Line" (http://www.vicrailstations.com/Cudgewa/Cudgewa.html). www.vicrailstations.net.. Retrieved 2008-06-09.

End of the broad gauge, the break-of-gauge platform at Albury station

Further reading

- Turton, Keith W. (1973). *The North East Railway. (A Lineside Guide) Melbourne to Wodonga*. Australian Railway Historical Society (Victorian Division). ISBN 0858490137.
- John C. Jennings and Robert K. Whitehead (July 2005). *Seymour - A Railway Town*. Seymour and District Historical Society. ISBN 0 9751658 0 1.
- Turton, Keith W. (1973). *Six And A Half Inches From Destiny. The first hundred years of the Melbourne-Wodonga Railway 1873-1973.*. Australian Railway Historical Society (Victorian Division). ISBN 0 85849 012 9.
- Covell, C.M. (November 1967). "Some Impressions and Reflections of the North-Eastern Line". *Australian Railway Historical Society Bulletin*: 230–244

External links

- Statistics and detailed schematic map (http://www.vicsig.net/index.php?page=infrastructure&section=lineguide&line=Albury) at the vicsig (http://www.vicsig.net/) enthusiast website
- History Victoria: The North East Railway Story (http://www.historyvictoria.org.au/pdf/Wangaratta.pdf)

Rail_transport_in_Victoria

Railways of Victoria	
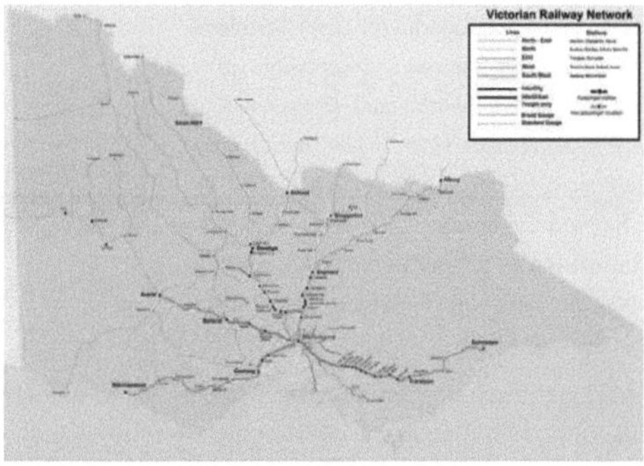 Victorian railway network, passenger lines in colour, freight only lines in grey	
Locale	Victoria, Australia
Dates of operation	1854–
Track gauge	5 ft 3 in (1600 mm) broad gauge, 4 ft 8 $\frac{1}{2}$ in (1435 mm) standard, some 3 ft 6 in (1067 mm) and 2 ft 6 in (762 mm) narrow gauge
Length	1712 km passenger, 4129 km total.[1]

Rail transport in Victoria, Australia, is provided by a number of railway operators who operate over the government-owned railway lines. Victorian lines use 5 ft 3 in (1600 mm) broad gauge, with the exception of a number of standard gauge 4 ft 8 $\frac{1}{2}$ in (1435 mm) freight and interstate lines, a few experimental 2 ft 6 in (762 mm) narrow gauge lines, and various private logging, mining and industrial railways.

The extent of the state railway network in 1947

Railways were privately owned and operated, until the State Government established the vertically integrated Victorian Railways in 1883. This remained until corporatisation occurred in the 1980s, followed by privatisation in the 1990s. Passenger services today are operated by Metro Trains Melbourne in suburban Melbourne with electric multiple units, and V/Line in regional Victoria with diesel trains. Freight services are operated by Pacific National and other private operators such as El Zorro, Specialised Container Transport and QRNational.

The rail network radiates from the state capital, Melbourne, with two main interstate links to Sydney and Adelaide, as well as major lines running to regional centres,[2] upgraded as part of the Regional Fast Rail project. The rail network reached a peak in 1942 but steadily declined, as branch and cross country lines were closed up to the 1980s.[1]

Victoria does not have a dominant mining base as with other states, and has traditionally been more dependent on agriculture for rail freight traffic.[3] By the 1990s road transport had captured most general freight traffic,[1] with an average of only 6.1 million tonnes of intrastate freight carried each year between 1996 and 1998;[1] containers being the major cargo, followed by cement, logs, quarry products, and steel.

History

First lines

Australia's first steam operated railway was a $2\frac{1}{2}$ mi (4.0 km) Victorian broad gauge line between the Melbourne (or City) Terminus (on the site of modern day Flinders Street Station) and Sandridge (now Port Melbourne), constructed by the Melbourne and Hobson's Bay Railway Company and opened in September 1854.[4] The first country line in Victoria was from Melbourne to Geelong,[5] opened in 1857 by the Geelong and Melbourne Railway Company.[6] The early years saw the majority of lines being constructed by privately companies, this not changing until the 1870s when the Government Railway Department (Victorian Railways) had built their own lines.[2]

Pre 1910 Flinders Street Station building

The suburban network expanded to the east from Princes Bridge railway station to Richmond in 1859, then later to Brighton and Hawthorn by the early 1860s. the initial suburban lines were all built by different private companies centred upon Flinders St, being amalgamated into the Melbourne and Hobson's Bay United Railway Company by 1865,[2] with public ownership not coming until 1878.[7] In 1862 Victorian Railways lines had reached the great gold rush towns of Bendigo and Ballarat, and in 1864 railways were extended to the Murray River port of Echuca.[7]

In 1883 the first connection to another State's rail system was made, when the Albury-Wodonga line was completed to the New South Wales Government Railways' station at Albury, requiring a break-of-gauge to New South Wales' 4 ft 8 $\frac{1}{2}$ in (1435 mm) (standard gauge).[8] It was then followed in 1887 by a connection with the broad gauge South Australian Railways at Serviceton, with the *Intercolonial Express* (now *The Overland*) to Adelaide running between the capitals.[4] Additional trunk lines were also built though the 1870s, with rails extended to Sale, Portland and Colac; and the first branch lines built.[2] It was a time of improved train safety, with the first interlocking of railway signalling to protect trains provided in 1874,[7] and tests of continuous train brakes carried out in 1884.[7]

In 1884 State Parliament passed *The Railway Construction Act*, which authorised fifty-nine new lines to almost every corner of the state, and thus became known as the *Octopus Act*. The proposed lines would serve both new agricultural communities and support suburban land speculation.[2] [9] It was also this decade that the first narrow gauge line was opened from Wangaratta to Whitfield, with three other lines following by 1910.[7] The South Gippsland line was also opened from Dandenong to Leongatha by 1891.[8] However by the late 1890s the majority of the state was now covered in railways, with the exception of the Mallee country in the north west of the state which saw further line openings,[2] such as the Mildura line in 1903.[8]

A new century

The *Spirit of Progress* headed by locomotive S301 *Sir Thomas Mitchell* near Kilmore East in 1938. (Photo: The late John L. Buckland)

In 1907 the A2 class steam locomotives were introduced, 185 in number they were Victorian Railway's main passenger locomotive until the arrival of the diesel electric B class in 1952.[7] On 20 April 1908 the Sunshine train disaster occurred, killing 44 passengers in the worst Victorian accident ever.[7]

Electrification of the Melbourne suburban network was also carried out, with the first train running in 1919.[10] By 1924 210,000 passengers passed through the main city terminal of Flinders Street Station per weekday.[7] St Kilda was Melbourne's busiest suburban station with 4½ million journeys, followed by Footscray with almost 4 million, then Elsternwick, Ascot Vale, Essendon and Balaclava.[7] Technology advancements continued, with 3 position automatic signals introduced in 1915,[11] the conversion of screw couplings to knuckle couplers from 1924,[12] and the first remote controlled signalling provided in 1925.[7] The first level crossing flashing lights were installed at Mentone on Moorabin Road in 1932.[7]

Despite the World War I, significant extensions of rail lines were carried out, particularly in the wheat growing areas in the north west and west. The new lengthy parallel lines were considered to be cheaper to operate that the numerous short spurs such as those in the Goulburn Valley.[2] By 1930 the railway map of Victoria was largely complete, with the best land settled and the remaining land marginal for agriculture, with a number of lines built across the state border into the Riverina of NSW.[2] On November 1937 the first run of the *Spirit of Progress* was made, a streamlined all air conditioned train run between Melbourne and Albury, led by the matching S class steam locomotives.[10]

The Victorian Railways biggest steam locomotive H 220 *Heavy Harry* entered service in 1941,[7] at a time when the railways were struggling with the needs of the war effort. In 1943 the Victorian Railways employed 25,450, had 577 steam locos and 12 electric locos on register along with 19,823 goods wagons and 1,499 passenger cars, running along 4758 miles (7657 km) of lines.[7] The network reached its largest extent in 1942, covering 7668 route kilometres.[1]

Post war rebuilding

Steam locomotive X 32, hauling a livestock train

After World War II the railways were run down, with Operation Phoenix unveiled in 1950, involving the expenditure of £80,000,000 over 10 years.[13] Works included electrification to Traralgon, new Harris suburban trains, the Walker railmotors, and approximately 3,000 new goods wagons.[7]

On 14 July 1952 the VR entered the diesel era, with the delivery of the first B class mainline locomotive,[13] with the commissioning of the first mainline electrification scheme in Australia in July 1954 to Warragul.[7] March 1954 saw Queen Elizabeth II tour Victoria by Royal Train, the first time a reigning monarch had travelled on the VR,[7] 1954 also saw the last steam locomotive to enter service, J class 559,[7] as well as the last four wheeled open wagons being built.[12] The fifties also saw the loss of a number of short branch lines, particularly country where the only traffic had been timber or livestock.[2]

In the 1960s the break of gauge at Albury was eliminated, with the opening of the North East standard gauge line in 1962.[13] The new line aided freight traffic between the state capitals, and enabled though passenger trains, such as the *Southern Aurora* and the *Intercapital Daylight.* At the same time the sixties was also the end of steam, with the demolition of the massive North Melbourne Locomotive Depot on 20 January 1961.[7]

The *Spirit of Progress* led by GM class locomotives in the 1970s

1965 saw the Victorian Railways produce at £193,727 surplus, but by 1973 it had turned into a $86,086,361 deficit. On 20 July 1976 the Laverton derailment occurred, killing one passenger, in what was the last railway passenger fatality not involving a road vehicle.[7] By the late seventies roadside goods and country railmotor services had been replaced by road transport, and branch lines outside the grain producing areas were now virtually non-existent.[2] The Lonie Report delivered in 1980 recommended the closure of all country passenger service except that to Geelong, elimination of a number of suburban railways, and moving small-volume freight from rail to road.

New deal

The 1980s saw corporatisation of the Victorian Railways carried out, with the railway commissioners replaced by VicRail and later government authorities. New liveries on trains were unveiled, as elderly *"red rattlers"* were replaced by new trains.

Train exiting the Melbourne underground loop

1981 saw the Melbourne underground loop open in January, followed by the new air conditioned Comeng suburban trains and "N" type country passenger carriages in September the same year.[7] Country services were also sped up under the *New Deal* by the closure of 35 of small wayside stations.[14] Country passenger services saw the last of the non-air-conditioned wooden bodied passenger cars withdrawn from service in 1986, replaced by new "H" set carriages. Trials were also carried out for further upgrades, with locomotive A85 re-geared for 160 km/h operation in a series of test runs between Glenorchy and Lubeck in the state's west in July 1986.[15]

N class locomotive with "N" type country passenger carriages.

It was also the end of an era, with freight trains having their guards vans and guards abolished from 1985, and the carriage of livestock ended in 1986.[7] The last run of the *Spirit of Progress* and *Southern Aurora* passenger trains were also made in 1986, on 3 August.[16] Working practices were also altered, with through working of C class locomotives introduced between Melbourne and Adelaide in 1982.[13] Previously Victorian locomotives were detached at the state borders, and replaced by locomotives from the next state. New locomotive were also introduced, with the G class and N classes enabling the retirement of many of the 1st generation diesels.[17]

Cuts continued to the rail network, with larger centralised silos in the north western area of the state, and replacement of traditional safeworking systems by systems that required no local staff saw further stations de-manned.[2]

Privatisation

The 1990s saw costs further reduced on the rail network. Guards were removed from passenger trains in 1989, and driver only suburban trains commenced running in 1993, with the last suburban train crewed by a guard running in November 1995.[7] In 1994 the National Rail Corporation was established, taking profitable steel, and intermodal trafficks from V/Line;[18] and the Melbourne to Adelaide standard gauge line was opened in 1995 moving the break of gauge.[19] Moves towards privatisation begun in 1997, with V/Line split into V/Line Passenger and V/Line Freight by Bayside Trains and Hillside Trains being separated from the Public Transport Corporation in 1998.[7] V/Line Freight was sold to Rail America, known as Freight Victoria in 1999, followed by National Express taking over Bayside Trains and V/Line Passenger in 2000, with Connex Melbourne taking over Hillside Trains.

End of private lease

Steve Bracks negotiated a premature end to the "under-rail" Victorian country Broad Gauge network lease just minutes before the caretaker mode began before the state election of November 2006. This cost around $125 million.

In May 2008 John Brumby arranged a 45 year lease to the ARTC of the single track Seymour-Albury section. Part of the negotiation involved Victoria contributing money for the track to be upgraded and standardised and for Wodonga to be bypassed.

Today, the state consists of four networks: the electrified metropolitan system operated by Metro Trains Melbourne, the country passenger network operated by V/Line and upgraded as part of the Regional Fast Rail project, the standard gauge interstate lines to Adelaide and Sydney, and the grain network in the north west of the state, connected to the ports at Geelong and Portland.[2]

Infrastructure

Network route kilometres (January 2007)*(excludes interstate and metropolitan passenger lines)*[1]

	Broad gauge	Standard gauge	Total
Passenger and freight	1712	nil	1712
Freight-only	1673	368	2041
Services suspended	290	86	376
Total	3675	454	**4129**

The trunk railway lines of Victoria are double track, some built as such, and others duplicated later on. Early sections of the suburban network were double track, with later additions being single track that were later duplicated.

The Bendigo and Geelong–Ballarat mainlines were both built as double track in the 1860s, but were singled in the 1930s and first decade of the 21st century respectively. The north-east line to Seymour was duplicated in the mid 1880s[20] and remains so today, the Gippsland line to Moe was duplicated in the 1950s in conjunction with increased briquette traffic,[21] and the busiest country line in the state to Geelong was progressively duplicated from 1959 to 1981.[6]

Few railway tunnels exist in Victoria, with the exception of the Melbourne City Loop. The longest tunnel before the opening of the loop was on the Fyansford Cement Works Railway (near Geelong), where a 1300 metre long tunnel existed on a narrow gauge quarry railway.[22] Of those open today the longest is the single track 422 metre long Geelong Tunnel,[23] followed by the double track 385 metre Elphinstone and the 390 metre long Big Hill Tunnels on the Bendigo line.[24] A 154 metre long tunnel also exists on the Healesville line, as well as three tunnels on the suburban Hurstbridge line, and another on the freight lines under Footscray station.[25]

On the interstate railway corridors there are independent sections of track due to the use of both broad and standard gauges in the state, these sections being from Melbourne to Seymour and on to Albury (constructed in the 1960s), and Melbourne to Geelong (completed in 1995). The gauge issue also sees dual gauge track used, in areas including Maryborough, North Geelong, and various freight terminals in Melbourne.

The maximum speed of suburban electric multiple units and locomotive hauled trains is 115 km/h, with diesel multiple units permitted up to 130 km/h on the same track, and up to 160 km/h on specified lines.[26] The maximum axle loading of freight wagons is 20 tonnes,[27] with locomotives of up to 22 tonne axle loading operating.[26] Train lengths are limited to 1200 metres, except on the main interstate lines where 1500 metre long trains are permitted.[27]

Track gauge

Dual gauge (broad and standard) track work

The majority of the railways of Victoria are of 5 ft 3 in (1600 mm) broad gauge, presenting break-of-gauge difficulties when connections were made with New South Wales which uses the 4 ft 8 $\frac{1}{2}$ in (1435 mm) standard gauge. In addition, the Victorian Railways experimented with four short narrow gauge lines of 2 ft 6 in (762 mm) in the early 20th century. Efforts to eliminate the gauge issue were proposed many times in the intervening years, with a Royal Commission in 1921 deciding "*that the gauge of 4-ft. 8.5-in. be adopted as the standard for Australia; that no mechanical, third rail, or other device would meet the situation, and that uniformity could be secured by one means only, viz., by conversion of the gauges other than 4-ft. 8.5-in.*"[28]

By the 1950s, interstate traffic was suffering from the break-of-gauge at the New South Wales state border, and a parallel standard gauge line was opened from the Melbourne to join the New South Wales system in 1962, along with a bogie exchange depot to allow wagons to operate across the broad and standard gauge networks.[1] The second interstate link from Victoria to Adelaide (the oldest single-gauge inter-capital line dating from 1887) was converted from broad to standard gauge in 1995 enabling rail traffic from Victoria to access the rest of the nation without disruption.[1]

Today the standard gauge network consists of the two main interstate lines, and a number of branch lines in the far west of the state.[29] Gauge conversion of 2000 kilometres of track was announced in May 2001 by the Victorian Government under the Linking Victoria program, but did not proceed due to the difficulty of achieving any agreement with then track manager, Freight Australia.[1] The works would have covered 13 lines,[30] including the Mildura line via Geelong, Ballarat, and Maryborough; the north western Victorian grain network; and the lines centred upon Seymour and Benalla in the north east. Conversion of the regional and suburban passenger networks is not envisioned.[30] In 2008 the conversion of the North East line was announced, with the conversion of 200 kilometres (120 mi) of broad gauge track to standard gauge between Seymour and Albury providing double track along the section.[31]

Loading gauge

The Victorian loading gauge for vehicles is sized between that of British and American practices. Wagons may be up to 22.85 metres long, 2.97 metres wide, and carry loads up to 4.27 metres above the rail height.[27] Double stacking of container wagons is not possible under these limits, with 2655 mm (8 ft 8.5 in) high containers the largest permitted, with the exception of 3200 mm (10 ft 6 in) high containers on some routes.[27] Trials were made with the 4D double deck passenger train on some suburban lines, requiring alterations to overhead bridges and structures, but no double deck trains are used today.

Signalling

Compare to Australian railway signalling

Victoria uses a mix of railway signalling practices: British route signalling with home and distant signals (*2 position signalling*) and American speed signalling (*3 position signalling*).[32]

Mechanical *2 position* semaphore signalling at North Geelong

Semaphore signals were used on the very first railway lines, but only a bare minimum were provided as the time interval system being relied upon instead.[33] The first interlocking of signals to protect trains was provided in 1874,[7] as before this time conflicting moves could be made. The design of the signals also progressed, with the *disc* type siding signals first introduced in 1885,[34] and the lower quadrant *somersault* type main line signals adopted in 1887,[32] both of which are still in use today. Green was not adopted as the *All Right* colour until 1898, with white being used before this time.[34] Red was the usual colour of all signal arms, until yellow was chosen as the colour for distant signals in 1926, with full adoption made in 1930.[34] Colour light signals first appeared in 1918, and by 1924 they were the standard for new installations.[32]

Mechanical interlocking frame for signal control at Avenel

The safeworking of trains between stations on the early lines was time interval working,[33] where a train would be allowed to leave a given time after the train before it. With heavier traffic this method became unsafe, with Staff and Ticket working on single lines adopted from 1873, and telegraph block working from 1878 on double lines.[35] Both of these systems ensured that only one train would be in a section of track at one time. Telegraphic block working was then replaced with *Winters Block* working between 1883 and 1888,[36] a system that is a predecessor of the *Double line Block* system which is still used today. Later years saw variations made to the *Staff and Ticket* system, with busier lines provided with Electric Staff working which provided greater safely when more trains ran.

Modern LED colour light *3 position* signalling

Heavier suburban traffic on the Melbourne network saw a greater strain on the block working then used, which required a large number of manned signal boxes to enable trains to run close together.[37] As a result it was decided to adopt power signalling under the Automatic Block System (ABS) of safeworking, where the presence of trains automatically control the signals after them, providing a safe distance between trains. Introduced from 1915, the system was based on American speed signalling practice with GRS2A upper quadrant mechanical signals with two arms able to indicate up to 5 different speed aspects to train drivers.[32] These signals were later replaced by colour light signals which are the standard today, but the old mechanical style remained until 2001.[38]

A variant of the *Automatic Block System, Automatic and Track Control* (ATC) has since been introduced, which provides the same benefits as ABS on single lines of track, while still ensuring only one train in a section at a time. Centralised Traffic Control was also introduced in the 1960s on the new standard gauge line to Albury,[20] and then on the main interstate line to Adelaide, allowing trains to be directed from a distance.

Today little mechanical signalling remains, with local signal boxes controlling signals abolished from many areas as part of the Regional Fast Rail project.[32] Today the suburban network and busier regional lines use variants of Automatic Block Signalling, while quieter lines use the Train Staff and Ticket or Train Order systems of safeworking.[39] Train protection has also progressed, with the Train Protection & Warning System also introduced on major passenger lines as part of the Regional Fast Rail project.[32]

Rollingstock

Rollingstock in Victoria has used air braking since tests of continuous train brakes were carried out in 1884.[7] Initially screw couplings were used to connect wagons, but from 1924 automatic knuckle couplers were introduced, with buffers removed from freight wagons by 1960.[12]

Locomotives

Preserved R class steam locomotive

In Victoria groups of locomotives of the same design are classed together by letter, in a system introduced in 1886.[40] Initially low letters were given to passenger classes, and high letters for goods classes, but from about 1916 this pattern was discarded.[40] Within classes locomotives are individually numbered, in a continuous number block which is unique to the class. In some cases locomotives have been renumbered to keep number blocks continuous, and in some cases there is no correlation between the number blocks used for the steam and diesel locomotives of the same class letter.[41] In addition, some classes start the number series with a *0*, while others start with a *1*.[41]

The first locomotives used in the state were small steam locomotives, and by the end of the 19th century 0-6-0 tender engines were used on goods, and 4-4-0 configurations for passenger workings.[42] The majority of locomotives were imported from Britain, from companies such as Beyer, Peacock and Company, Robert Stephenson and Company, R and W Hawthorn and George England.[42] The Williamstown Workshops also built locomotives locally, as did the Phoenix Foundry in Ballarat.[42]

As the size of train increased, so did the locomotives. The 4-6-0 wheel arrangement became popular for passenger and mixed traffic work with the D3 and A2 classes, and the 2-8-0 arrangement on goods with the K, J and C classes lasting into the 1960s.[42] These were then followed by the 4-6-2 S class *Pacifics* for the *Spirit of Progress* express, and the single 4-8-4 H class locomotive H220, the biggest Victorian Railways steam locomotive.[43] The final steam locomotives built for the Victorian Railways were the 4-6-4 R class and 2-8-0 J classes. Although they were of more modern design, their advantages were overshadowed by the simultaneous arrival of the first diesels.[40]

Electric locomotives were first acquired with the electrification of the suburban railways, the E class suburban engines acquired in 1923 and 1928–29, followed by the L class from 1953 when the mainline to Traralgon was electrified for briquette traffic.[40]

Dieselisation occurred from 1951 with the F class shunter, but the B and S classes of 1952 and 1957 revolutionised main line operations.[40] They were then followed by the T and Y classes in 1955 and 1963 which displaced steam from the branch lines and yards.[40] Apart from the F class, Clyde Engineering has had a monopoly on Victorian diesel-electric locos.[44] as the Australian licensee of General Motors EMD engines and traction motors, fitting them into locally designed bodies. By 1980s the first generation locomotives were approaching the end of their lives, with the electric locomotives withdrawn by 1988,[40] and the modern N and G classes allowing the withdrawal of lower power T and Y classes.[45]

Today the former Victorian Railways locomotive fleet has been split into two, with the N, A and P classes utilised by V/Line on passenger services, with the remainder with Pacific National or other private operators in freight use.[44]

No new passenger locomotives have been built since the 1980s, with Diesel Multiple Units being acquired instead. Freight operations have seen the re-powering of the G and X classes, as well as the restoration to service of stored locomotives, now up to 50 year old.[46]

Passenger

Modern V/Line VLocity diesel train purchased for the Regional Fast Rail project

Early passenger services were operated with 4 and 6 wheeled "dogbox" passenger carriage with small compartments and no side corridors. Later years saw the provision of side corridors to provided access down the train, and onboard toilets. Sleeping carriages was first introduced in 1887,[25] and dining cars from 1908.[47]

Larger bogie rollingstock started to appear from the turn of the 20th century,[47] with the E type carriage stock introduced on *The Overland*,[48] and smaller W type carriage stock on intrastate trains. The first air conditioned carriage was introduced in 1935, when one of the E type carriages was fitted.[49] All steel carriages came to the Victorian Railways in 1927, with the construction of the "Avoca" and "Hopkins" dining cars,[50] followed by the S type carriage stock for the new *Spirit of Progress* in 1937.[10]

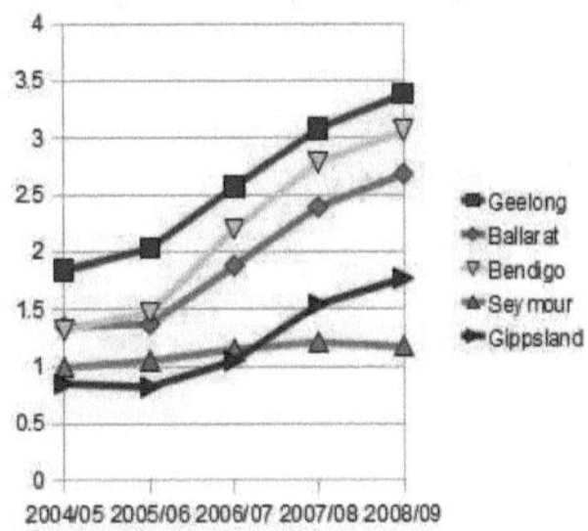

V/Line passenger figures (in million passenger trips) for major rail routes since 2004

On the Melbourne suburban network Electric Multiple Units were introduced speeding up services. Experiments were also made with various diesel and petrol railcars for use smaller branch lines, with the DERM being the most successful, remaining in service from 1928 to 1991. A fleet of Walker railcars was also introduced in the 1950s, along with Z type saloon carriage stock[47] for both intra and interstate trains.

By the 1980s country passenger services were run down, and older wooden rolling stock was now approaching their use by date. As a result, the N type carriages were introduced from 1981,[47] followed by the converted H type stock from 1984.[47] Since then, diesel multiple units have become the norm for new purchases, with the Sprinter introduced in 1992, and the VLocity from 2005.

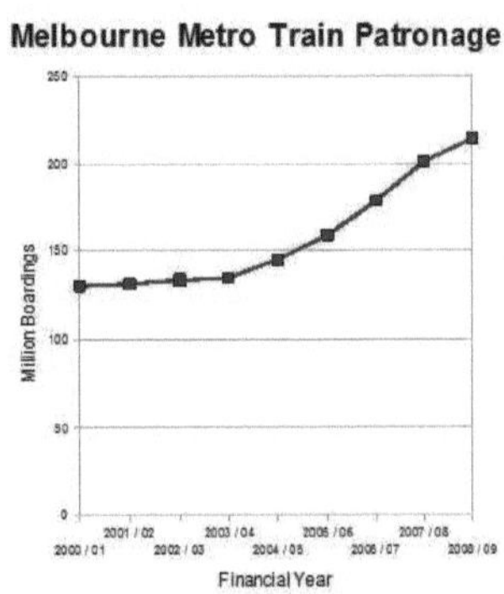

Metropolitan train patronage 2000–2009 based on official state government figures.

Freight

Various Victorian bogie wagons

Early wagons were built on four wheeled under frames, but from 1871 bogie vehicles begun to appear.[18] These early wagons were of all wood construction, but later types had wood bogies on steel underframes, then followed by all steel wagons.[12] Groups of wagons of the same design are classed together by a multiple character alphabetical code, initially in a system restricted to Victoria only, but from 1979 the Railways of Australia four letter coding was introduced.[18] Under this system the first letter represent owner of the wagon, the second represents the general type of wagon, the third separates different classes of the same general type, and the four letter indicates the maximum running speed.[18]

The last four wheeled open wagons were built in 1958,[12] but were not scrapped in large numbers until the 1980s when new bogie wagons replaced them.[51] In 1987 the bogie wagon fleet numbered 5000, with approximately 700 grain hoppers, 800 container flats, 1000 louvred vans, 700 open wagons, 400 tank cars, and 300 flat wagons.[51] Today the broad gauge intrastate fleet numbers 2600,[12] with large numbers of louvred vans, open and flat wagons, and tank cars stored or scrapped due to the transfer of traffic to road following years of apathy by Governments.

Operations

Commodities

Traffic by year (by tonnes consigned)

Commodity	Percentage (1987)[3]	Percentage (1996 to 1998)[1]
Bulk grain (wheat, barley, oats)	31.4%	50.8%
Containers	10.6%	19.6%
Freight forwarders	9.5%	?
Mining, quarry products	7.9%	14.7%
Cement	6.8%	4.9%
Iron and steel	6.3%	?
Petroleum	4.4%	3.2%

Victoria does not have a dominant mining base as with other states, and has traditionally been more dependent on agriculture for rail freight traffic.[3] In the early 20th century rail was a "common carrier" and was required to carry almost any freight offered. After World War I road competition increased, until in 1933 legislation was passed to regulate trucks competing with rail on specific routes.[1] From 1974 to the 1980s intrastate road freight was deregulated, and rail "common carrier" obligations were removed, resulting in the loss of to road of much non-bulk freight.[1] In 1987 10.51 million tonnes of freight was carried by rail, with bulk grain being the main commodity consisting of 31.4% of traffic.[3]

Continued drought, road competition, and rationalisation of intrastate freight operations have seen this change in recent years, with an average of only 6.1 million tonnes of intrastate freight carried from each year between 1996 to 1998;[1] and containers now being the major traffic, with residual cement, logs, quarry, and steel trains operating. By the 1990s road transport has captured most general freight traffic,[1] with less than carload traffic was discontinued in July 2007 with the end of the *Fastrack* freight service.[52] An exception to the decline is interstate operations, with intermodal container traffic growing due to competition between private operators.

Companies

The first railways in Victoria were operated by private companies, but when these companies failed or defaulted, Victorian Railways took over. The Victorian Railways was initially known as the "Department of Railways" from the first appointment of staff in 1856, and on 1 November 1883 the Victorian Railways Commissioners Act established the "Victorian Railways".[53]

Pacific National operated freight train

This situation remained until May 1973 when the Railways (Amendment) Act 1972 passed the management of the Railways from the Victorian Railways Commissioners to a Victorian Railways Board,[13] with rebranding as VicRail following in 1976.

Heritage train in the Victorian Railways livery

In 1983 VicRail was split into the State Transport Authority taking responsibility for the provision of country rail and road, passenger and freight services under the V/Line brand; and the Metropolitan Transit Authority taking over suburban passenger operations.[13] These authorities were then merged into the Public Transport Corporation in 1989, with the suburban services rebranded as "The Met".

In 1993 the Kennett Government offered long-distance country rail services previously run by the government-owned operator V/Line to private operators. While several rail services were replaced by road coach services, West Coast Railway successfully tendered to operate the railway line to Warrnambool, becoming the first private operator in Victoria.[54] In 1993 Professor Fred Hilmer presented the findings of the National Competition Policy Review Committee, known as the Hilmer Report, leading to the introduction of National Competition Policy in 1995. This allowed new rail freight operators to establish operations in Victoria. Specialised Container Transport commenced operating trains to Perth in 1995,[19] and Great Northern Rail Services started intrastate operations.

Privatisation of the Public Transport Corporation commenced under the Kennett Government of the 1990s. V/Line was split into separate freight and passenger divisions, and "The Met" was divided into "Hillside Trains" and "Bayside Trains". These separate bodies were all sold separately in 1999: V/Line Freight and the rural intrastate network to Freight Victoria; Hillside Trains and track lease to Connex Melbourne; and V/Line Passenger and Bayside Trains (with track lease) to the National Express Group.[13] Control of V/Line's former interstate rail freight infrastructure was passed to the federal Australian Rail Track Corporation. Competition in the rail freight industry has also seen the emergence of small operators such as El Zorro and Southern Shorthaul Railroad.

On 22 December 2002 National Express withdrew from their operations in Victoria and the government taking control temporarily,[55] V/Line reverted to government ownership on 1 October 2003,[56] and M-Train was taken over by Connex on 18 April 2004.[57] 16 August 2004 saw the Freight Victoria business and rural track lease was purchased by Pacific National,[58] but by November 2006 they entered into an agreement to sell the track lease back to the Victorian Government for $133.8 million,[59] with the sale completed on 7 May 2007 and V/Line becoming track manager.[60]

In December 2007 Pacific National announced plans to sell or close its grain transport and Portlink rural container business operations in Victoria.[61] The decision has been criticised as it will force grain growers to use higher cost road transport to transport the annual grain harvest from rural silos to the ports.[62] The decision has seen many commentators accuse Pacific National of only acquiring the operations of Freight Australia in 2004 for the purposes of asset stripping and eliminating competition in rail freight.[63] [64] In 2008 El Zorro has taken over the Warrnambool - Melbourne container service from Pacific National,[65] and moved into broad gauge grain services, while QRNational has taken over the Melbourne - Horsham container service.[66]

By line

Victoria's first railway was a 4 km Victorian broad gauge line between the Melbourne (or City) Terminus (on the site of modern day Flinders Street Station) and Sandridge (now Port Melbourne), constructed by the Melbourne and Hobson's Bay Railway Company and opened in September 1854. Today, **Melbourne's suburban railway network** consists of 16 electrified lines, the central City Loop subway, and 200 stations, with a total length of 372 km of the electrified lines, operated by Metro Trains Melbourne under franchise to the Government of Victoria.

In Gippsland the **Orbost line** was constructed as an extension of suburban lines between 1877 and 1888 to Bairnsdale, and extended eastwards to Orbost in 1916.[8] The line between Dandenong and Traralgon was electrified in 1954, but electrification was cut back progressively to Pakenham between 1987 and 2001. V/Line passenger trains now run as far as Bairnsdale.

The **South Gippsland line** was opened from Dandenong to Cranbourne, Leongatha and Port Albert between 1888 and 1892.[8] A number of branch lines were also built, with almost all closed in stages between the early 1970s and 1994, except for part has since been electrified and re-opened as part of the suburban network, and between Nyora and Leongatha where the South Gippsland Tourist Railway operates heritage services.

The **North East line** originated from a suburban line to Essendon in 1860, being extended to Wodonga by 1873, connecting with the New South Wales Government Railways at Albury at a break-of-gauge in 1883.[8] A standard gauge track was completed parallel with the broad gauge track from Albury to Melbourne in 1962.[13] V/Line services operate along the broad gauge line, while the Melbourne-Sydney Countrylink XPT runs on the standard gauge.

The **Shepparton line** was built from Mangalore to Shepparton in 1880 and extended to the New South Wales Government Railways at Tocumwal at a break-of-gauge in 1908.[8] V/Line passenger service run as far as Shepparton.

The **Bendigo line** was completed in 1862, with extensions were opened to Echuca in 1864, and Swan Hill in 1890,[8] followed by a number of lines throughout the north-west corner of the state. Only these lines continue to see V/Line services, with other lines in the region only seeing freight traffic.

The **main western line** had its beginnings in the first railway line from Melbourne to Ballarat, the Geelong-Ballarat line that opened in 1862.[6] The line extended from Ballarat to Ararat between 1874 and 1875,[8] but it was not until 1889 that the direct line between Melbourne and Ballarat was opened, built from both ends in stages until they met at Ballan. Further branch lines followed to Portland and other western towns. The line formed the first interstate railway line in Australia, when it was extended to the South Australian Railways at Serviceton in 1887. Up until the 1990s the Ballarat line was on the main route between Melbourne and Adelaide, as well as *The Overland* services to Adelaide, until the One Nation Program rerouted the main interstate line via Geelong and Maroona as standard gauge.[1]

The **Port Fairy line** was started when the Geelong and Melbourne Railway Company opened the Geelong line to Newport in 1857, being extended to Spencer Street Station in 1859.[6] The line was taken over by Victorian Railways in 1860 and a line was opened from Geelong to Ballarat in 1862,[6] and later extended south-east from 1876, reaching Warrnambool and Port Fairy in 1890.[6] Branch lines also existed to Queenscliff, Beech Forest, and a number of other towns.[6]

Private railways

Trains on the Yallourn 900mm Railway

A small number of private railways have also existed at various times. Some, such as the Geelong and Melbourne Railway Company, and the Melbourne and Suburban Railway Company, Melbourne and Hobson's Bay Railway Company, Melbourne and Essendon Railway Company and St Kilda and Brighton Railway Company were acquired by the Victorian Railways and made part of the core state network in 1860 and 1878 respectively. Other country branch lines were also built by private companies: namely the Kerang-Koondrook Tramway and the Deniliquin and Moama Railway,[67] both not being acquired until the 20th century.

In addition to the main Victorian rail network of the Victorian Railways and successors, a number of narrow gauge private railways and tramways have also existed for logging and mining purposes. These included the Yallourn 900mm Railway in the Latrobe Valley open cut coal mines, the Fyansford Cement Works Railway near Geelong, the Tyers Valley Tramway at Mount Baw Baw, and the Powelltown Tramway from Yarra Junction.

Most logging tramways operated in the Otway Ranges, Gippsland,[68] and the inner east of the Great Dividing Range; primarily between the 1850s and the 1950s, with only one surviving into the 1960s.[68] They were primarily of 3 ft 6 in (1067 mm) or 3 ft (914 mm) gauge, with 2 ft (610 mm), 2 ft 6 in (762 mm), 4 ft (1219 mm), 5 ft 3 in (1600 mm) and variants also used.[68]

Railway preservation

Organised railway preservation commenced in Victoria with the formation of the Puffing Billy Preservation Society in 1955, and operating under the Emerald Tourist Railway Board from 1977.[69] Formed to operate the narrow gauge 2 ft 6 in (762 mm) gauge railway in the Dandenong Ranges near Melbourne, the group continues to operate the railway today.

The demise of the last of the steam locomotives in Victoria commenced in the 1960s, with the Australian Railway Historical Society and Association of Railway Enthusiasts working with the Victorian Railways to have a number of locomotives preserved for the future.[70] In 1962 the ARHS Railway Museum was established at North Williamstown to house static exhibits,[71] and Steamrail Victoria was formed in 1965 to assist in the restoration of locomotives and carriages for use on special trains.[72]

By the 1980s a number of heritage railways had been established in Victoria on closed branch lines. These railways serve both as tourist attractions, and to preserve the railway past. The work of railway preservation groups has since expanded to retired railway carriages, electric multiple units, rail motors, and diesel locomotives. In 2006 heritage railways carried 542,000 patrons over 161 km of track; with 28 operational steam locomotives, 47 diesels, 14 railmotors, and 192 carriages.[73]

Heritage Railways and Operators include:

- Puffing Billy Railway, Belgrave
- Bellarine Peninsula Railway, Queenscliff
- Daylesford Spa Country Railway, Daylesford
- Victorian Goldfields Railway, Maldon
- Mornington Railway Preservation Society, Mornington
- Seymour Railway Heritage Centre, Seymour
- Steamrail Victoria, Newport

- Walhalla Goldfields Railway, Walhalla
- Yarra Valley Tourist Railway, Healesville
- South Gippsland Railway, Leongatha
- Melbourne Tramway Museum, Bylands

Rail trails

Victoria's tourist railways
Bellarine Peninsula Railway
Daylesford Spa Country Railway
Mornington Railway
Puffing Billy Railway
South Gippsland Railway
Victorian Goldfields Railway
Walhalla Goldfields Railway
Yarra Valley Tourist Railway

A number of former rail lines in Melbourne and regional Victoria have been converted into *rail trails*, for walking, cycling and horse riding. These make excellent tracks for beginners as the lines were originally designed to avoid steep gradients. Most lines are still in public ownership.[74] Some sections of the track are specially leased to neighbouring farmers for stock grazing. This reduces maintenance costs.

Trails around Melbourne include:[75]

- Inner Circle (Rushall to Royal Park)
- Outer Circle (Fairfield to Hughesdale)
- Rosstown (Hughesdale to Elsternwick)
- Hawthorn to Kew
- Red Hill (Merricks to Red Hill)

Trails around Victoria include:

- East Gippsland (Bairnsdale to Newmerralla)
- The Beechy (Colac to Beech Forest)
- Crater to Coast (Camperdown to Timboon)
- Ballarat–Skipton
- Murray to the Mountains (Wangaratta to Beechworth and Bright)
- Goulburn River High Country (Bonnie Doon to Mansfield)
- Great Southern Rail Trail (Leongatha–Foster, future extension to Yarram)
- Bass Coast Rail Trail (Anderson–Wonthaggi, future extension to Nyora)

Planned trails include:

- Warrnambool to Port Fairy

Legislation, governance and access

Key statutes

The prime rail statute in Victoria is the Transport Integration Act. The Act establishes the Department of Transport as the integration agency for Victoria's transport system. The Act also establishes and sets the charters of the State agencies charged with providing public transport rail services and managing network access for freight services, namely the Director of Public Transport and V/Line. In addition, the Act creates VicTrack which owns the public rail network and associated infrastructure. Another important statute is the Rail Management Act 1996[76] which confers powers on rail operators and provides for a rail access scheme for the State's rail network.

Safety

Regulation

The safety of rail operations in Victoria is regulated by the Rail Safety Act 2006 which applies to all commercial passenger and freight operations as well as tourist and heritage railways.[77] The Act creates a framework containing safety duties for all rail industry participants and requires rail operators who manage infrastructure and rolling stock to obtain accreditation prior to commencing operations. Accredited rail operators are also required to have a safety management system to guide their operations.

Sanctions applying to the safety scheme established under the Rail Safety Act are contained in the Transport (Compliance and Miscellaneous) Act 1983.[78] The safety regulator for the rail system in Victoria is the Director, Transport Safety (trading as Transport Safety Victoria) whose office is established under the Transport Integration Act 2010.

Investigation

Rail operators in Victoria can also be the subject of no blame investigations conducted by the Chief Investigator, Transport Safety or the Australian Transport Safety Bureau(ATSB). The Chief Investigator is charged by the Transport Integration Act[79] with conducting investigations into rail safety matters including incidents and trends. ATSB, on the other hand, has jurisdiction over the same matters where they occur on the Designated Interstate Rail Network.

Ticketing and conduct

Ticketing requirements for public transport in Victoria are mainly contained in the Transport (Ticketing) Regulations 2006[80] and the Victorian Fares and Ticketing Manual.[81] Rules about safe and fair conduct on trains and trams in Melbourne are generally contained in the Transport (Compliance and Miscellaneous) Act 1983[82] and the Transport (Conduct) Regulations 2005.[83]

Tourist and heritage railways

Tourist and Heritage Railways in Victoria are currently governed by provisions in the Transport (Compliance and Miscellaneous) Act 1983. In future, they will be regulated by the recently enacted Tourist and Heritage Railways Act 2010[84] which is yet to commence.

See also

- Railways in Melbourne
- Narrow gauge lines of the Victorian Railways
- Victorian broad gauge
- Dandenong railway line triplication
- Regional Rail Link

References

[1] John Hearsch (1 February 2007). "Victoria's Regional Railway Past, Present and Potential" (http://web.archive.org/web/20070830043728/http://www.rtsa.com.au/events/conferences/fffrr/text-for-rtsa-wagga-presentation-20070107.pdf) (PDF). *RTSA Regional Rail Symposium, Wagga Wagga*. Archived from the original (http://www.rtsa.com.au/events/conferences/fffrr/text-for-rtsa-wagga-presentation-20070107.pdf) on 30 August 2007. . Retrieved 2008-02-08.

[2] "Victorian Railway Maps 1860–2000" (http://www.vrhistory.com/VRMaps/). *Victorian Railways Resources*. Andrew Waugh. . Retrieved 2008-02-05.

[3] Peter Foote (October 1987). "V/Line Grain Handling". *Newsrail* (Australian Railway Historical Society (Victorian Division)) **15** (10): pages 292–297.

[4] "ARHS Railway Museum: History 1839–1900" (http://www.railwaymuseum.org.au/history.html). *ARHS Railway Museum*. . Retrieved 2008-02-05.

[5] ***The Greenwich Railway - 1857*** Clark, Allan Australian Railway History, June/July/August 2007 pp. 234–245; 260–279; 308–323

[6] "Geelong Line" (http://www.railgeelong.com/lineguide.php?line=geelong). *Rail Geelong*. Marcus Wong. . Retrieved 2008-02-05.

[7] "VR timeline" (http://www.victorianrailways.net/vr history/history.html). http://www.victorianrailways.net/. Mark Bau. . Retrieved 2008-02-05.

[8] Sid Brown (March 1990). "Tracks Across the State". *Newsrail* (Australian Railway Historical Society (Victorian Division)) **18** (3): pages 71–76.

[9] Venn, Michael (1976). "Speight, Richard (1838–1901)" (http://www.adb.online.anu.edu.au/biogs/A060189b.htm). *Australian Dictionary of Biography*. Canberra: Australian National University. . Retrieved 2008-02-17.

[10] "ARHS Railway Museum: History 1900–1950" (http://www.railwaymuseum.org.au/history2.html). *ARHS Railway Museum*. . Retrieved 2008-02-05.

[11] Fisher, Peter (2007). *Victorian Signalling: by Accident or Design?*. Australian Railway Historical Society (Victorian Division). ISBN 978 1 92089 250 0.

[12] Mark Bau. "Four wheeled open wagons of the Victorian Railways". *Notes from the Victorian Model Railway Society Prototype Modellers Meet 2007*.

[13] "ARHS Railway Museum: Victoria 1950 - now" (http://www.railwaymuseum.org.au/history3.html). *ARHS Railway Museum*. . Retrieved 2008-02-05.

[14] Scott Martin and Chris Banger (October 2006). "'New Deal' for County Passengers - 25 years on". *Newsrail* (Australian Railway Historical Society (Victorian Division)): page 319.

[15] "News". *Newsrail* (Australian Railway Historical Society (Victorian Division)) **14** (9): pages 24–27. September 1986.

[16] G. Carmichael (September 1986). "Aurora and Spirit of Progress depart". *Newsrail* (Australian Railway Heritage Society) **14** (9).

[17] Railmac Publications (1992). *Australian Fleetbooks: V/Line locomotives*. Kitchner Press. ISBN 0 949817 76 7.

[18] Norm Bray and Peter J Vincent (2006). *Bogie Freight Wagons of Victoria*. Brief History Books. ISBN 0 9775056 0 X.

[19] "ARTC History" (http://www.artc.com.au/Content.aspx?p=32). *Australian Rail Track Corporation*. . Retrieved 2008-02-05.

[20] Andrew Waugh. "Somerton" (http://www.vrhistory.com/Locations/Somerton.pdf) (PDF). *Victorian Station Histories*. . Retrieved 2008-02-08.

[21] S.E. Dornan and R.G. Henderson (1979). *Electric Railways of Victoria*. Australian Electric Traction Society. p. 91. ISBN 0 909459 06 1.

[22] "Fyansford Line" (http://www.railgeelong.com/lineguide.php?line=fyansford). *Rail Geelong*. Marcus Wong. . Retrieved 2008-02-08.

[23] "Geelong Tunnel" (http://www.railgeelong.com/locations.php?name=Geelong Tunnel). *Rail Geelong*. Marcus Wong. . Retrieved 2008-02-08.

[24] "Bendigo Line" (http://www.vicsig.net/index.php?page=infrastructure&line=Bendigo). *Vicsig Lineguides*. . Retrieved 2008-02-08.

[25] "The Railway Top 20!" (http://www.historyvictoria.org.au/pdf/questions.pdf) (PDF). *Royal Historical Society of Victoria*. . Retrieved 2008-02-05.

[26] "Addenda" (http://www.vline.com.au/pdf/rna/addenda.pdf) (PDF). *V/Line Network Service Plan* (http://www.vline.com.au/rna/rna/information_pack.html). V/Line Regional Network and Access. . Retrieved 2007-09-10.

[27] "Network Operating Requirements" (http://www.vline.com.au/pdf/networkaccess/networkserviceplan/network_operating_requirements.pdf/network_operating_requirements) (PDF). *V/Line Network Service Plan* (http://www.vline.com.au/about/networkaccess/infopack.html). V/Line Regional Network and Access. . Retrieved 30 June 2011.

[28] "Standardisation of Railway Gauges" (http://www.abs.gov.au/Ausstats/abs@.nsf/dc057c1016e548b4ca256c470025ff88/f7d0f9dda8f0a320ca2569de0028b414!OpenDocument). *Year Book Australia, 1967*. Australian Bureau of Statistics. 25/01/1967. . Retrieved 2008-02-08.

[29] Mark Bau. "RAILWAY MAP VICTORIAN LINES" (http://steam4me.railpage.org.au/files/vrmap.pdf) (PDF). . Retrieved 2005-10-01.

[30] Victorian Auditor General (August 2006). "Rail Gauge Standardisation Project" (http://www.audit.vic.gov.au/reports_mp_psa/psa1603.html). *Victorian Auditor General's Office*. . Retrieved 2008-02-08.

[31] "$500m rail link upgrade for Victoria" (http://news.ninemsn.com.au/article.aspx?id=571907). news.ninemsn.com.au. . Retrieved 2008-05-30.

[32] Fisher, Peter (2007). *Victorian Signalling: by Accident or Design?*. Australian Railway Historical Society (Victorian Division). p. *x*. ISBN 978 1 92089 250 0.

[33] Fisher 2007, pp. 39

[34] Fisher 2007, pp. 62

[35] Fisher 2007, pp. 64

[36] Fisher 2007, pp. 66

[37] S.E. Dornan and R.G. Henderson (1979). *Electric Railways of Victoria*. Australian Electric Traction Society. p. 20. ISBN 0 909459 06 1.

[38] Fisher 2007, pp. 89

[39] "Safeworking in Victoria" (http://www.vicsig.net/index.php?page=infrastructure&section=safeworking). *Vicsig*. . Retrieved 2008-02-08.

[40] Railmac Publications (1992). *Australian Fleetbooks: V/Line locomotives*. Kitchner Press. p. 3. ISBN 0 949817 76 7.

[41] Examination of all classes in VicRail Public Relations (1981). *Power Parade*. Melbourne, Vic.: VicRail Public Relations Division. ISBN 0 7241 3323 2.

[42] Railmac Publications (1992). *Australian Fleetbooks: V/Line locomotives*. Kitchner Press. p. 2. ISBN 0 949817 76 7.

[43] "What to see - H220" (http://www.railwaymuseum.org.au/whattosee_h220.html). *ARHS Railway Museum*. . Retrieved 2008-02-05.

[44] "VR - V/Line - VLP/ FA Locomotives" (http://locopage.railpage.org.au/vline.html). *Locopage*. . Retrieved 2008-02-05.

[45] Railmac Publications (1992). *Australian Fleetbooks: V/Line locomotives*. Kitchner Press. p. 12. ISBN 0 949817 76 7.

[46] "Mainline Locomotives" (http://www.cfcla.com.au/mainline.htm). *Chicago Freight Car Leasing Australia*. . Retrieved 2008-02-05.

[47] "Victorian Railways - Loco hauled Passenger Carriages" (http://www.pjv101.net/cd/pages/type_car.htm). *Peter J. Vincent's Website*. . Retrieved 2008-02-08.

[48] "V&SAR Joint Stock Sleeping Cars" (http://www.pjv101.net/cd/pages/c191m.htm). *Peter J. Vincent's Website*. . Retrieved 2008-02-08.

[49] Mark Bau. "AE 1st class wooden compartment pass. cars" (http://www.victorianrailways.net/pass cars/pass car pages/ae.html). . Retrieved 2008-02-08.

[50] Mark Bau. "Avoca dining car" (http://www.victorianrailways.net/pass cars/pass car pages/avoca.html). . Retrieved 2008-02-12.

[51] "V/Line Freight Rollingstock Fleet - 1 July 1987". *Newsrail* (Australian Railway Historical Society (Victorian Division)) **15** (10): page 303. July 1987.

[52] "Goodbye to Victorian Louver Vans" (http://www.railpage.com.au/f-t11329739.htm). *Railpage Australia forums*. . Retrieved 2008-02-08.

[53] "ARHS Railway Museum: Victoria 1839–1900" (http://www.railwaymuseum.org.au/history.html). *ARHS Railway Museum*. . Retrieved 2008-02-05.

[54] Philip Hopkins (2004-07-19). "End of the line for West Coast Rail" (http://www.theage.com.au/articles/2004/07/18/1090089033538.html?from=storylhs). *The Age*. . Retrieved 2008-02-05.

[55] Richard Web (14 March 2004). "The long goodbye" (http://www.theage.com.au/articles/2004/03/13/1078594612786.html). *The Age*. . Retrieved 2008-02-17.

[56] "V/Line Background" (http://www.vline.com.au/about/news/en/12/157/article.aspx). *V/Line*. . Retrieved 2008-02-05.

[57] Selma Milovanovic, Rachel Wells (17 April 2004). "Down Frankston way it's a popular line" (http://www.theage.com.au/articles/2004/04/16/1082055646029.html?from=storyrhs). *The Age*. . Retrieved 2008-02-17.

[58] "SALE OF FREIGHT AUSTRALIA TO PACIFIC NATIONAL" (http://www.dpc.vic.gov.au/domino/Web_Notes/newmedia.nsf/b0222c68d27626e2ca256c8c001a3d2d/a9042091d2782e61ca256ef2007af392!OpenDocument). *OFFICE OF THE PREMIER (Media Release)*. 16 August 2004. . Retrieved 2008-02-17.

[59] "Toll sells rail lease". Canberra Times. 2 November 2006. pp. 17.

[60] "RAIL BUY BACK DEAL COMPLETE" (http://www.legislation.vic.gov.au/domino/Web_Notes/newmedia.nsf/b0222c68d27626e2ca256c8c001a3d2d/6f0846798c60e597ca2572d4007d9d99!OpenDocument). *MINISTER FOR PUBLIC TRANSPORT (Media Release)*. 7 May 2007. . Retrieved 2008-02-05.

[61] "Pacific National to sell or close rail freight business" (http://www.abc.net.au/news/stories/2007/12/12/2116312.htm). *ABC News website*. 12 December 2007. . Retrieved 2007-12-17.

[62] Easdown, Geoff (12 December 2007). "Asciano delivers grain haulage shock to farmers" (http://www.news.com.au/heraldsun/story/0,21985,22909741-664,00.html). *Herald and Weekly Times website*. . Retrieved 2007-12-17.

[63] "Loco Lines" (http://www.rtbuvicloco.com.au/locolines/LL 2 07.pdf). *Newsletter for Locomotive Enginemen of the Rail, Tram & Bus Union – Victoria*. May 2007. . Retrieved 2007-12-17.

[64] "Freight Australia Exiting" (http://www.railpage.com.au/f-t11336862-0-asc-s0.htm). *Railpage Australia forums*. 12 December 2007. . Retrieved 2007-12-17.

[65] "Rail freight deal New operator avoids closure" (http://warrnambool.yourguide.com.au/news/local/general/rail-freight-deal-new-operator-avoids-closure/1198056.html). *The Warrnambool Standard*. warrnambool.yourguide.com.au. 7 March 2008. . Retrieved 2008-04-04.

[66] "Rail to roll on" (http://wimmera.yourguide.com.au/news/local/news/general/rail-to-roll-on/1229118.aspx). *Wimmera Mail Times*. yourguide.com.au. 28 July 2008. . Retrieved 2008-07-30.

[67] Australian Bureau of Statistics - Year Book Australia, 1920 (http://abs.gov.au/AUSSTATS/abs@.nsf/b4005c38619c665aca25709000203b8d/73d48ba2d3c1bb9bca2569de0028b417!OpenDocument)

[68] Frank Stamford (14–15 April 2001). "Australian and New Zealand Logging Tramways - Differences from North American Practice" (http://www.lrrsa.org.au/log_diff.pdf) (PDF). . Retrieved 2007-12-22.

[69] "Preservation" (http://web.archive.org/web/20061008232730/http://www.puffingbilly.com.au/fan/heritage/histpbps.htm). *Puffing Billy*. Archived from the original (http://www.puffingbilly.com.au/fan/heritage/histpbps.htm) on 8 October 2006. . Retrieved 2008-02-19.

[70] *Excursion to Fyansford*. A.R.H.S. (Victoria Div). 20 April 1965.

[71] "About us" (http://www.railwaymuseum.org.au/aboutus.html). *ARHS Railway Museum*. . Retrieved 2008-02-19.

[72] "About Us" (http://web.archive.org/web/20080319000244/http://www.steamrail.com.au/srv_aboutus.html). *Steamrail Victoria*. Archived from the original (http://www.steamrail.com.au/srv_aboutus.html) on 19 March 2008. . Retrieved 2008-02-19.

[73] "Statistics for Heritage and Tourist Railways in Victoria" (http://www.atr.org.au/statistics.htm). *Association of Tourist Railways Inc.*. . Retrieved 2008-02-19.

[74] Department of Sustainability and Environment. "Victoria's Rail Trails" (http://www.dse.vic.gov.au/DSE/nrenrt.nsf/LinkView/D81DE59D8CBF32444A2567E70026134002D27D699EC3FC914A256B890018E95D). www.dse.vic.gov.au. . Retrieved 2010-06-03.

[75] "Railtrails Australia Home" (http://www.railtrails.org.au). . Retrieved 2009-01-20.

[76] Rail Management Act 1996 (http://www.legislation.vic.gov.au/Domino/Web_Notes/LDMS/LTObject_Store/LTObjSt3.nsf/DDE300B846EED9C7CA257616000A3571/742548D962C75039CA257761002C52FD/$FILE/96-79a050.pdf)

[77] Official copy of the Rail Safety Act from the Victorian Government legislation web site - http://www.legislation.vic.gov.au/Domino/Web_Notes/LDMS/LTObject_Store/LTObjSt3.nsf/DDE300B846EED9C7CA257616000A3571/6D4190C62DE6207FCA257761002C568A/$FILE/06-9a016.pdf

[78] See Part 7 of the Act. Official copy of the Transport (Compliance and Miscellaneous) Act 1983 from the official Victorian Government legislation site - http://www.legislation.vic.gov.au/Domino/Web_Notes/LDMS/LTObject_Store/LTObjSt5.nsf/DDE300B846EED9C7CA257616000A3571/7BB774D3E0245B77CA2577CE00030B90/$FILE/83-9921a153.pdf

[79] Transport Integration Act 2010, Part 7.

[80] Transport (Ticketing) Regulations 2006 (http://www.legislation.vic.gov.au/Domino/Web_Notes/LDMS/LTObject_Store/LTObjSt5.nsf/DDE300B846EED9C7CA257616000A3571/C6D777772BB02CD3CA2577A5007CB77B/$FILE/06-86sr004.pdf)

[81] Victorian Fares and Ticketing Manual - Metlink - Your guide to public transport in Melbourne and Victoria (http://www.metlinkmelbourne.com.au/fares-tickets/victorian-fares-and-ticketing-manual)

[82] Victorian Law Today Act (http://www.legislation.vic.gov.au/Domino/Web_Notes/LDMS/PubLawToday.nsf/a12f6f60fbd56800ca256de500201e54/6ff43affd338ea35ca2577c60082b59c!OpenDocument)

[83] Victorian Law Today Statutory Rule (http://www.legislation.vic.gov.au/Domino/Web_Notes/LDMS/PubLawToday.nsf/b12e276826f7c27fca256de50022686b/2e01dcbed2e5778cca2574e20017da2c!OpenDocument)

[84] Victorian Statute Book Act (http://www.legislation.vic.gov.au/Domino/Web_Notes/LDMS/PubStatbook.nsf/f932b66241ecf1b7ca256e92000e23be/e2cbeabed816bbcdca2577c1001d6f7a!OpenDocument)

Bibliogrpahy

- Fisher, Peter (2007). *Victorian Signalling: by Accident or Design?*. Australian Railway Historical Society (Victorian Division). ISBN 978 1 92089 250 0.

External links

History:

- Mark Bau's Victorian Railways (http://www.victorianrailways.net) - History of rolling stock, locomotives, and signalling
- Peter J. Vincent (http://www.pjv101.net/) - History of rolling stock and wagons of the Victorian Railways
- Victorian Railways Resources (http://www.vrhistory.com/) - Network maps and analysis at ten year intervals, and histories of stations
- Signal Diagrams (http://signaldiagramsandphotos.com/) - Digitised copies of Victorian Railways and V/Line signalling diagrams
- Rail Geelong (http://www.railgeelong.com/) - History of the railways of the Geelong region

Today:

- Victorian Railway Stations (http://www.vicrailstations.com/) - photographs of stations of the Victorian network
- Vicsig (http://www.vicsig.net) - Information on present infrastructure, operations, signalling, and locomotives
- V/Linecars (http://www.vlinecars.com/) - V/Line carriages and operations website

Serviceton_railway_line

Serviceton railway line, Victoria	
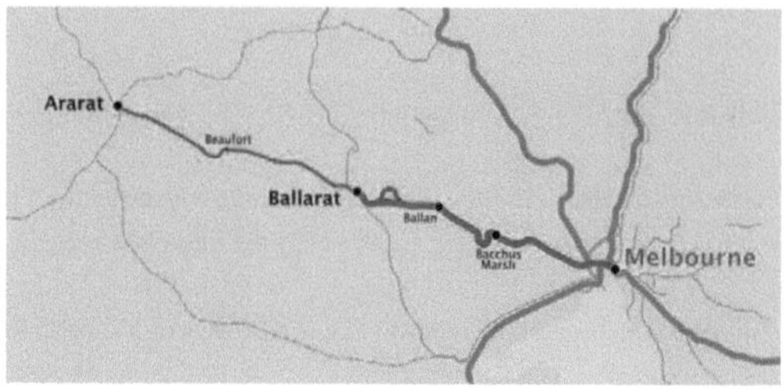	
Line details	
Opened	1862
Completed	to South Australian border in 1887
Tracks	Single 1600 mm (5 ft 3 in) (broad gauge) to Ararat, 1435 mm (4 ft 8 $^1/_2$ in) (standard gauge) beyond
Used by	V/Line passenger services to Ballarat and Ararat, freight and *The Overland* beyond
Connections	Geelong-Ballarat, Mildura, Portland, Avoca, Hopetoun and Yaapeet lines
Former connections	Buninyong, Skipton, Bolangum and Carpolac lines
Rail transport in Victoria	
List of stations **Closed railways**	

The **Serviceton railway line** (also known as the **Western line**) is a railway serving the west of Victoria, Australia. The line links the state capital of Melbourne to the cities of Ballarat and Ararat, and once extended to the South Australian border as part of the Melbourne-Adelaide railway. In this role it has been replaced by the Western standard gauge line.

Services

Metro Trains Melbourne operates suburban passenger train services along the inner section of the line as far as Sunshine as the Sydenham Line, while V/Line services operate as the Interurban Ballarat Line and the Intercity Ararat Line. During peak hour some services originate and terminate at Bacchus Marsh. Passengers on the inner section of the line to Melton are permitted to use Metcard tickets to access the services, with this section marked as the Melton line on suburban network maps. Beyond Ararat the line is part of the Western standard gauge line and sees *The Overland* to *Adelaide.*

Freight services also use the line between Ballarat and Melbourne, coming off the Mildura line. The Western standard gauge also sees a large number of interstate freight trains as part of the Melbourne-Adelaide railway.

History

Serviceton railway station at the state border

The first railway line from Melbourne to Ballarat was via Geelong. Construction of the Geelong-Ballarat line began in 1858 and took nearly four years to complete. The project employed 3,000 men and cost approximately 1.5 million pounds. The line officially opened in 1862, with the first train running, at a rate of 15 miles per hour, on 10 April taking around four and a half hours to reach Geelong, amid various complications - inability to pull up a hill and the lack of firewood. The return trip commenced at 12:20pm, arriving in Ballarat at 3pm.[1]

The line was extended from Ballarat to Beaufort in 1874 and to Ararat in 1875. It took until 1889 for a direct line from Melbourne to Ballarat to be opened, in part due to the difficult terrain between Bacchus Marsh and Ballan.

Up until 1995 the line through Ballarat was the main route for freight trains between Melbourne and Adelaide, as well as *The Overland* services to Adelaide, and V/Line services to Horsham and Dimboola. V/Line passenger rail services beyond Ballarat were withdrawn on 27 May 1994,[2] and in 1995 the One Nation Program rerouted the main interstate line via North Shore and Maroona and the broad gauge line between Ballarat and Ararat was closed to all traffic. The broad gauge passenger service from Ballarat to Ararat was not reinstated until 2004.

Work began in 2005 on upgrades between Sunshine and Ballarat as part of the Regional Fast Rail project, which saw heavier tracks and concrete sleepers installed, renewal of ballast and a new signalling system. A number of deviations were built to ease curves on the line, as well as seven new bridges, with the Lal Lal Creek and Moorabool River bridges amongst the longest in Australia.[3] The works were completed and the first VLocity train ran on the line on 22 December 2005.[4]

As part of the Transit Cities program [5], a new station was opened at Wendouree on 14 June 2009.[6] [7]

In June 2008 Pacific National and grain handler GrainCorp announced they would cease rail freight operations on the Yaapeet to Dimboola line. The decision was based on the Rail Freight Network Review chaired by former Deputy Prime Minister Tim Fischer, which placed the Yaapeet line at the lowest level of investment priority. As a result farmers at Rainbow, Jeparit and Yaapeet must truck their grain to GrainCorp's Warracknabeal grain centre instead of the Rainbow depot. A local council attacked the plan as it would result in the need to upgraded roads in the area.[8]

Branch lines and extensions

Ballarat region

V/Line VLocity train at the junction with the Melbourne-Ballarat line at Warrenheip station

A branch line was built from Ballarat to Newtown and Skipton in 1883. This line closed in 1985. A branch was built between Newtown, Cressy and Irrewarra (east of Colac) about 1910. This line was closed in 1953.

A branch line was built from Linton junction to the Ballarat Cattle Yards and Redan in 1886. This line was closed in 1997.

Ararat and west

Clocktower at Ballarat station

The line was extended from Ballarat to Beaufort in 1874 and Ararat in 1875, Stawell in 1876, Murtoa in 1878, Horsham in 1882, Dimboola in 1882 and linked with the South Australian Railways at Serviceton in 1887.

Until 1993, two daily broad gauge V/Line trains ran from Melbourne to Dimboola as well as the nightly *The Overland* service from Melbourne to Adelaide. Under the Kennett Government's economic reforms, the V/Line trains were then cut back to Ararat. The Ballarat to Ararat section was closed in 1994 with gauge standardisation of the line, which made reinstatement of a V/Line service beyond Ararat virtually impossible.

The Overland service was re-routed via Maroona after the works were complete, while the broad gauge passenger service from Ballarat to Ararat was not reinstated until 2004 by the Labour State Government.

A branch line was built from Stawell to Grampians station in the 1880s, but it was closed in the 1930s.

A branch line was built from Lubeck (between Stawell and Murtoa to Rupanyup in the 1890 and extended to Marnoo by 1910 and Bolangum in the 1920s. It was closed in 1983.

A branch line was opened from Murtoa to Warracknabeal in 1886 and extended to Beulah in 1893, Roseberry and Hopetoun in 1994 and Patchewollock in the 1920s. The section from Hopetoun to Patchewollock was closed in the 1980s. The line from Murtoa to Hopetoun was converted to standard gauge in 1995. It has not carry passengers for many years and grain services between Roseberry to Hopetoun were suspended in 2005.

A branch line was completed between Horsham, East Natimuk and Noradjuha in the 1880s and extended to Balmoral by 1920. This was connected to Cavendish and Hamilton in the 1920s. The line from Noradjuha to Hamilton was closed in the 1970s. A branch was built from East Natimuk to Carpolac in 1894. This line and the branch to Noradjuha were closed in the 1990s.

A branch was opened from Dimboola to Jeparit in 1894 and extended to Rainbow in 1899 and Yaapeet in 1914. The line was converted to standard gauge in 1995. The line was deemed un-useable after severe flooding in January 2011. An anouncement was made in September 2011 by the Victorian Goernment, that they would contribute $5.3 million to restoring the line from Dimboola to Rainbow. A branch line was opened from Jeparit to Yanac by 1920, which was closed in 1986.

Ararat southwards

A line was built from Ararat to Portland in 1877, via Maroona, Hamilton, and Heywood. A number of branch lines were built from this line, none of which are in use today. This line was converted to standard gauge in 1995.

A line was opened from Gheringhap on the Geelong - Ballarat line to Cressy and Maroona in 1913. This was converted to standard gauge in 1995 and became the Western standard gauge line between Melbourne and Perth.

Line guide

Broad Gauge tracks are in red, and Standard Gauge is in blue.

References

[1] "Ballarat Australia History - Rail Heritage" (http://www.ballarathistory.org/transrlh.html). Ballarat History Central. .

[2] VictorianRailways.net - VR Timeline (http://www.victorianrailways.net/vr history/history.html)

[3] "Transport planning and projects - Regional Fast Rail" (http://www.doi.vic.gov.au/doi/internet/planningprojects.nsf/headingpagesdisplay/regional+fast+rail). State Government of Victoria, Department of Infrastructure. 15 May 2006. .

[4] V/Line - First Day of VLocity Revenue Operation (Ballarat) (http://wongm.fotopic.net/c802268.html)

[5] http://www.dse.vic.gov.au/DSE/nrenpl.nsf/LinkView/7EF2CA1E91EB87C3CA256D480003CEFFCB3A095B12F1DCC1CA256D190029AA6B

[6] "Public transport - Regional railway station projects" (http://www.doi.vic.gov.au/Doi/Internet/transport.nsf/AllDocs/5B8CEA11A3D8F676CA2571F800022DAD?OpenDocument#ballarat). State Government of Victoria, Department of Infrastructure. 21 September 2007. .

[7] "'Wendouree' to be new station name" (http://ballarat.yourguide.com.au/detail.asp?story_id=1040283). The Courier. 21 August 2007. .

[8] "Council slams rail line plans" (http://wimmera.yourguide.com.au/news/local/news/general/council-slams-rail-line-plans/793956.aspx). *Wimmera Mail Times*. yourguide.com.au. 2008-06-20. . Retrieved 2008-07-02.

External links

- Statistics and detailed schematic map (http://www.vicsig.net/index.php?page=infrastructure&section=lineguide&line=Ararat) at Vicsig

Orbost_railway_line

Orbost railway line, Victoria	
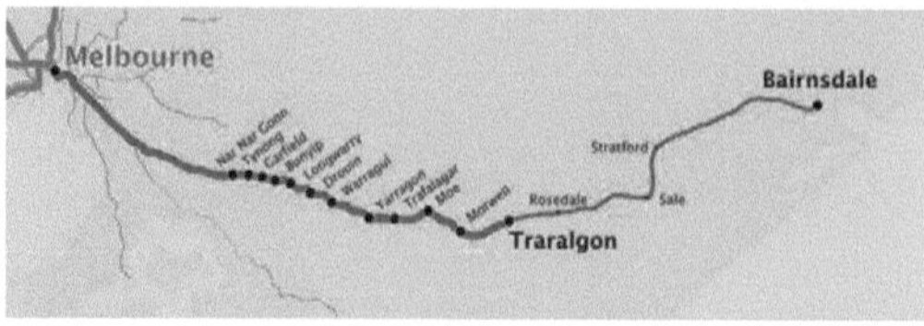	
Line details	
Commenced	1877
Completed	1916
Closed	1987[1]
Used by	V/Line passenger services, freight traffic
Rail transport in Victoria	
List of stations **Closed railways**	

The **Orbost railway line** (also known as the **Gippsland line**) is a railway serving the Latrobe Valley and Gippsland regions of Victoria, Australia. Running east from the state capital Melbourne though the cities of Moe, Morwell, Traralgon, Sale and Bairnsdale until reaching the eventual terminus of Orbost.

Today the line ends at Bairnsdale, with the remaining section to Orbost now existing as the East Gippsland Rail Trail, a shared bicycle, walking, and horseriding track.[2]

Services

Metro Trains Melbourne operates suburban passenger services along the inner section of the line as the Pakenham Line, while V/Line services operate as the Interurban Traralgon Line and the Intercity Bairnsdale Line. Freight services also use the line, operated by Pacific National.

History

The current end of the line at Bairnsdale

Rail lines were built to Gippsland in the 1870s and initially played a crucial role in developing agricultural industries in Gippsland as well as tourism. It also played a crucial role in the development of coal mining in the Latrobe Valley in the 1920s. At its peak, the railway travelled as far east as Orbost and there are still frequent services to many of the towns. Some of the disused rail lines have been turned into tourist railways and/or rail trails.

The Melbourne and Suburban Railway Company opened a line from Princes Bridge railway station to Punt Road (Richmond) and South

Yarra in 1859, Prahran in 1859 and Windsor in 1860, connecting with the St Kilda and Brighton Railway Company line. This line was extended to Dandenong, Pakenham, Warragul, Moe, Morwell, Traralgon, Sale, Stratford and Bairnsdale between 1877 and 1879. It was extended to Orbost in 1916. The line east of Sale was closed in 1994, but was reopened to Bairnsdale in 2004.

Section of line at Morwell Loop upgraded as part of the Regional Fast Rail project

The railway to Orbost opened in 1916 and operated until 1987, principally carrying timber and farming produce. In the early days of the railway's operation dedicated passenger trains ran but these ceased by the 1930s. The track infrastructure was dismantled in 1994. The line traversed a mixture of farmland, hills and heavily forested country. It included numerous bridges, including the Stoney Creek Trestle Bridge, the largest of its kind in Victoria.

In 1954 the line beyond Dandenong was electrified mainly because of the expected briquette traffic from the brown coal mines in the Latrobe Valley, and over the next two years most of the line between Dandenong and Pakenham was duplicated and provided with power signalling, although Narre Warren to Berwick was not done until 1962. In due course, the rail transport of briquettes petered out as industry converted to natural gas and homes were converted to other forms of heating.

Electrification was cut back to Warragul in 1987, when suburban services started to Warragul. Electrification and suburban services were cut back to Bunyip in 1998, before ceasing beyond Pakenham in 2001. The Victorian Regional Fast Rail project upgraded one of the two lines in 2005, between Pakenham and Traralgon.

Branch lines

A branch line was built north from Warragul in stages from the 1890s, reaching Noojee in 1919. This was closed in stages from 1954 to 1958 [3].

The private SECV Morwell Interconnecting Railway bridge crossing the main Gippsland railway

The narrow gauge (762 mm) Walhalla branch line was completed across mountainous country from Moe to Erica and Walhalla in 1910. The Platina to Walhalla section closed in 1944, Erica to Platina in 1952, and finally Moe to Erica in 1954 [4]. The northern-most section between Thomson and Walhalla stations has been reopened as a tourist railway by the Walhalla Goldfields Railway, and provides regularly scheduled trains.

A branch line was also opened from Moe to Thorpdale in 1888, which was closed in 1958.

The Yallourn branch was opened from Hernes Oak (between Moe and Morwell) to Yallourn in 1922 to serve the adjacent power station development. It was replaced by a line from Moe to Yallourn in 1953 because its route was required for brown coal mining,[5] but the new line closed in 1987, having been disused since the late 1970s.[5] [6]

The Mirboo North branch line was opened in stages from Morwell to Mirboo North between 1885 and 1886, however was closed in 1974.[7] The route of the line was partly dug up as part of the Hazelwood open cut mine. The Maryvale paper siding also connects to the main line at Morwell and remains open today for regular freight traffic.

The loop line via Maffra was opened from Traralgon to Heyfield, Maffra and Stratford in 1887 and closed in stages between 1987 and 1993. A branch line was opened from Maffra to Briagolong in 1889 and closed in 1952.

Several timber tramways also existed from many of the stations between Pakenham and Yarragon.

Significance

The expansion of the railway in the late 1870s helped to develop Gippsland. It enabled milk from western Gippsland to be sold fresh into Melbourne while the dairy industry of East Gippsland provided cheese and butter. It also enabled development of west Gippsland's market gardening and orcharding industry for sale in Melbourne markets.

It also encouraged the development of a tourism industry notably at Lakes Entrance. It did however end coastal shipping traffic and the use of Sale and Bairnsdale as ports.

In the 1920s, the Gippsland railway played an important role in developing the mining of lignite coal and the development of the Latrobe Valley for power generation primarily serving Melbourne and Victoria. This saw the development of industry in towns such as Yallourn, Morwell, Traralgon, Moe, Warragul and Drouin.

The development of the Gippsland Railway helped fuel the Melbourne land boom in the 1870s. The original departure point for the railway was Oakleigh with the line connecting Oakleigh and Melbourne not built until 1879. The Victorian railways bought land in Oakleigh for use as workshops. Oakleigh became a centre of what was known as "railway fever" as developers developed and marketed houses close to rail lines between Oakleigh and other suburbs for use of workers travelling to and from their job. At the height of the land boom in 1888, land sales were being held two or three times a week in the district. The collapse of the land boom in 1889 eventually contributed to banking collapses in 1893 and the major depression of the 1890s.

The Gippsland railway remains a significant passenger corridor on the V/Line network although its use for freight business has now declined to only one major customer, being Australian Paper's export traffic from Maryvale

Tourist railways and rail trails

Many of the lines in Gippsland have closed because they had become uneconomic. Some of these have been turned into tourist railways including:

- the Walhalla Goldfields Railway between Moe and Walhalla; and
- the South Gippsland Railway between Leongatha, Korumburra and Nyora.

Walhalla Goldfields Railway

Other stretches of line have become rail trails for use by bicyclists. These include:

- the Noojee Trestle Bridge Trail in the upper reaches of the Latrobe River;
- a section between Erica and the Thomson River on the former Walhalla line (pending eventual reconstruction of the Walhalla Goldfields Railway's tourist line between Thomson and Erica stations).;
- Collins Siding to Tyers Valley Rail Trail near Moe;
- Moe to Yallourn Rail Trail;
- Mirboo North to Boolarra Rail Trail;
- Great Southern Rail Trail from Leongatha to Foster;
- Bass Coast Rail Trail Anderson to Wonthaggi;
- East Gippsland Rail Trail from Bairnsdale to Nowa Nowa, and in 2005 being extended to Orbost; and the
- Gippsland Plains Rail Trail from Stratford to Maffra.

Line guide

Bold stations are termini.

Preserved substation and overhead wiring at Bunyip

An old section of the bridge across the Avon River near Stratford

References

[1] "Rail » Infrastructure » Orbost Line" (http://www.vicsig.net/infrastructure/line/Orbost). *Vicsig Railways site* (http://www.vicsig.net/index.php). VICSIG. 1997-2009. . Retrieved 2009-11-08.

[2] East Gippsland Shire Council (2007). "East Gippsland Rail Trail". James Yeates Printing & Design.

[3] Fiddian 1997, p. 170-171.

[4] Fiddian 1997, p. 169-170.

[5] S.E. Dornan and R.G. Henderson (1979). *Electric Railways of Victoria*. Australian Electric Traction Society. p. 88. ISBN 0 909459 06 1.

[6] "VR History" (http://www.victorianrailways.net/vr history/history.html). www.victorianrailways.net. . Retrieved 2008-06-18.

[7] "Railways of the Latrobe Valley" (http://www.latrobe.vic.gov.au/webfiles/media/link/march03/link mar 03 - page 07.pdf). Latrobe City (http://www.latrobe.vic.gov.au). . Retrieved 2006-06-07.

Further reading

- Fiddian, M (1997). "Trains, Tracks, Travellers".

External links

- South Gippsland Railway (http://www.sgr.org.au)
- Statistics and detailed schematic map (http://www.vicsig.net/index.php?page=infrastructure&section=lineguide&line=Bairnsdale) at the vicsig (http://www.vicsig.net/) enthusiast website

History

- City of Monash page on the Gippsland Railway (http://www.monash.vic.gov.au/history/section-a-4.html)
- Gippsland Rail Trails (http://www.railtrails.org.au/states/vic/gippsland.htm)
- Monash University gazeteer on Oakleigh (http://www.arts.monash.edu.au/ncas/multimedia/gazetteer/list/oakleigh.html)
- Monash University gazeteer on Moe (http://www.arts.monash.edu.au/ncas/multimedia/gazetteer/list/moe.html)

Portland_railway_line

Portland railway railway line, Victoria	
Line details	
Opened	1877
Rail transport in Victoria	
List of stations Closed railways	

The **Portland railway line** is a railway line in south-western Victoria, Australia. It runs from the main Western standard gauge line at Maroona through Hamilton to the port town of Portland.

History

The line was built as 1600 mm (5 ft 3 in) (broad gauge) from Ararat to Maroona, Hamilton, Heywood and Portland in 1877.[1] It was converted to standard gauge in 1995.[2]

Freight operator Pacific National indefinitely suspended all rail services to the town of Portland in 2004, impacting local companies including Portland Aluminium, transport company Kalari, and freight broker Anchor Logistics. Portland container traffic was conveyed between Maroona and Portland on grain services twice a week, but Pacific National said that due to the drought there were no trains to attach the loading to. On the route the price differential between rail and road is $12.97 per tonne in rail's favour.[3] Pacific National closed their Portland operations in March 2008, with GrainCorp leasing a limited number of locomotives and rolling stock from them, but favour transporting grain to the Port of Geelong instead.[4]

In July 2008 the Victorian Government announced a $15 million upgrade to the line, raising maximum speeds to 80 kilometres an hour as they were previously. Australian Rail Track Corporation signed a 50 year lease to manage the line and will start the upgrade within months.[5] The line was finally transferred from V/Line to ARTC on 22 March 2009.[6]

In September 2008 it was announced that freight traffic would restart using the line, with operator El Zorro signing a multi-million deal with miner Iluka Resources to carry containerised mineral sands from Portland in the south-west to Melbourne, with Iluka saying rail transport was cheaper than road.[7] The train continues to run today on an irregular basis.[8] In March 2010 the State Government announced a $3.97 million grant for the construction of a rail loading facility alongside the Iluka Resources mineral sand separation plant (near Hamilton railway station) for the rail transport of heavy mineral concentrate extracted from the Iluka mine site at Ouyen via the standard gauge Murtoa - Hopetoun railway.[9]

Grain trains on the line restarted with the 2010 harvest, with AWB Limited and their rail partner El Zorro using the GrainCorp terminal to load 26,250 tonnes of canola bound for Pakistan on 4 March 2010.[10]

Branch lines

A short lived branch line was opened from Dunkeld to Penshurst in 1890 but closed only eight years later.[1]

A branch line was opened from Hamilton north to Cavendish between 1910 and 1920. This was connected to the Horsham - Balmoral railway in the 1920s, which itself connected back to the main Serviceton line. This line was closed south of Noradjuha, (just south of East Natimuk) in the 1970s.

A branch line from Hamilton north-west to Coleraine was opened in 1882, being closed in 1977.[1]

A branch line was opened between Hamilton and Koroit in the 1890, where it connected to the South West line via Warrnambool, but was closed in 1977.[1]

A branch line was opened from Branxholme north-west to Casterton in 1884. This line was closed in 1977.[1]

A branch line was opened from Heywood west to Mount Gambier in 1917.[1] Service was suspended in 1995 due to the standardisation of the Maroona - Portland line. There are regular calls for the Heywood - Mount Gambier line to be standardised.[11] [12] [13]

Services

The only regular service on the line is the thrice weekly El Zorro containerised mineral sands train from Portland.[7] The last passenger train between Ararat and Portland was on 12 September 1981, operated by a DRC railcar.[14] A new passenger station at Portland had been officially opened on 29 June 1968.[15]

References

[1] Sid Brown (March 1990). "Tracks Across the State". *Newsrail* (Australian Railway Historical Society (Victorian Division)): pages 71–76.

[2] "VICSIG - Infrastructure - Line Data Portland line" (http://www.vicsig.net/index.php?page=infrastructure&line=Portland). www.vicsig.net.. Retrieved 2008-10-14.

[3] Bill Meldrum (17 December 2004). "Freight off the rails Pacific National suspends rail services to Portland" (http://web.archive.org/web/20050124083914/http://www.spec.com.au/display.asp?Id=4613). *Portland Observer*. web.archive.org. Archived from the original (http://www.spec.com.au/display.asp?Id=4613) on 2005-01-24.. Retrieved 2008-07-02. (via Internet Archive)

[4] Bill Meldrum (30 June 2008). "State loses focus on Portland line" (http://www.spec.com.au/index.asp?sp=2&id=8879). *Portland Observer*. www.spec.com.au.. Retrieved 2008-07-02.

[5] "Rail track upgrade announced for Portland" (http://www.abc.net.au/news/stories/2008/07/16/2305366.htm). *ABC News*. www.abc.net.au. 16 July 2008.. Retrieved 2008-07-16.

[6] V/Line Weekly Operational Notice No. 11/2009

[7] "Rail freight back on Portland-Maroona line" (http://www.abc.net.au/news/stories/2008/09/03/2353873.htm). *ABC News*. www.abc.net.au. 3 September 2008.. Retrieved 2008-09-04.

[8] "Wongm's Rail Gallery - Mineral sands train" (http://wongm.railgeelong.com/mineral-sands-train). wongm.railgeelong.com.. Retrieved 2010-03-18.

[9] Premier of Victoria (18 March 2010). "Media Release: GOVERNMENT HELPS PUT MINERAL INDUSTRY ON TRACK" (http://www.premier.vic.gov.au/newsroom/9805.html). www.premier.vic.gov.au.. Retrieved 2010-03-18.

[10] Rob McKay (March 4, 2010). "Shipment breaks Portland grain export drought" (http://www.lloydslistdcn.com.au/archive/2010/march/4/canola-shipment-breaks-portland-grain-export-drought). *Lloyd's List DCN*. www.lloydslistdcn.com.au.. Retrieved 2010-03-18.

[11] "New push for Mount Gambier railway" (http://www.borderwatch.com.au/archives/401). *The Border Watch*. www.borderwatch.com.au. 17 July 2008.. Retrieved 2008-10-14.

[12] "Penola rail pledge confirmed" (http://www.borderwatch.com.au/archives/415). *The Border Watch*. www.borderwatch.com.au. 18 July 2008.. Retrieved 2008-10-14.

[13] BILL MELDRUM (13 February 2008). "Portland railyards given the shunt" (http://www.spec.com.au/?sp=2&id=8400). *Portland Observer*. www.spec.com.au.. Retrieved 2008-10-14.

[14] Chris Banger (March 1997). "Rail Passenger Service Withdrawals Since 1960". *Newsrail* (Australian Railway Historical Society (Victorian Division)): pages 77–82.

[15] Vincent Adams Winter (1990). *VR and VicRail: 1962 - 1983*. p. page 205. ISBN 0 9582069 3 3.

External links

- Information about the line at Vicsig (http://www.vicsig.net/index.php?page=infrastructure&line=Portland)

Port_Fairy_railway_line

Port Fairy railway line, Victoria	
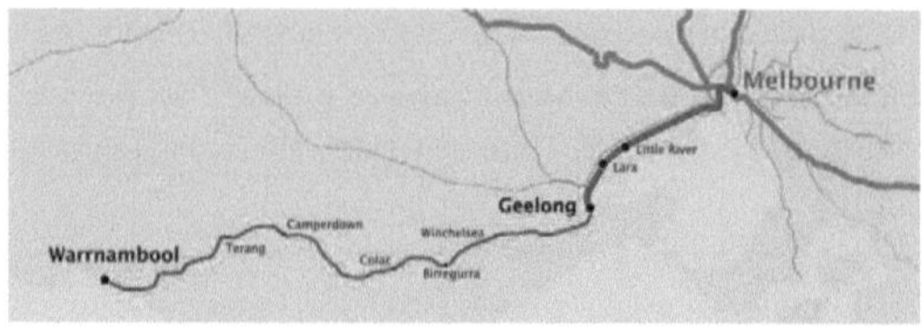	
Line details	
Opened	1857
Completed	1890
Tracks	Double broad gauge track to Geelong, single beyond. Single standard gauge to North Shore
Used by	V/Line passenger services, freight traffic
Connections	Western SG, Geelong - Ballarat
Former connections	Queenscliff, Geelong Racecourse, Wensleydale, Forrest, Cressy, Beech Forrest, Alvie, Timboon, Mortlake, Port Fairy, Hamilton lines
Rail transport in Victoria	
List of stations **Closed railways**	

The **Port Fairy railway line** (also known as the **South West line**) is a railway serving the south west of Victoria, Australia. Running from the state capital Melbourne though the cities of Geelong and Warrnambool, it once terminated at the coastal town of Port Fairy until truncated to Dennington (outside Warrnambool). The line continues to see both passenger and freight services today.

Services

Metro Trains Melbourne operates suburban passenger services along the inner section of the line as the Werribee Line, while V/Line services operate as the Interurban Geelong Line and the Intercity Warrnambool Line. For 11 years from 1993 until 31 August 2004 the Melbourne to Warrnambool services were run by the private West Coast Railway company.[1] Freight services also use the line, operated by Pacific National and now El Zorro.

History

The Geelong line was originally built by the private Geelong and Melbourne Railway Company and opened on 25 June 1857, the line being sold to the Victorian Railways in 1860.[2] The line was built as a single track, with the being progressively duplicated from the 1950s to the 1980s.[3]

The line was later extended south-west, to Winchelsea in 1876, Colac in 1877, Camperdown in 1883, Terang in 1877 and Warrnambool, Dennington and Port Fairy in 1890. This line is now closed beyond Warrnambool.[4] A branch line was also built from Koroit (between Warrnambool and Port Fairy), Penshurst and Hamilton (on the Ararat – Portland railway) in the 1890s. A connection was also opened between Penshurst and Dunkeld (also on the Ararat – Portland railway) in about 1900.

Branch lines also followed, from South Geelong to Queenscliff in 1879, both Moriac (between Geelong and Winchelsea) to Wensleydale and Terang to Mortlake in 1890, Birregurra (between Winchelsea and Colac) to Forrest in 1891, Camperdown to Timboon in 1892, Irrewarra (east of Colac) to Cressy in 1910, and Colac to Alvie in 1923.

A narrow gauge (762 mm (2 ft 6 in)) branch was also opened from Colac to Beech Forest in 1902 to carry timber, and extended to Weeaproniah and Crowes in 1911.

The branch lines begun to be cut back from the late 1940s, the Wensleydale line being the first in 1948, followed by the Forrest line in 1957, the Irrewarra line in 1953, the Alvie line in 1954, the Beech Forest (Crowes) line in 1962, and the Timboon line in 1988.

Line guide

Broad Gauge tracks are in red, and Standard Gauge is in blue.

The twin broad gauge and single standard gauge tracks

References

[1] Philip Hopkins, End of the line for West Coast Rail, The Age, July 19, 2004 (http://www.theage.com.au/articles/2004/07/18/1090089033538.html)

[2] "Rail Geelong - Geelong Line History" (http://railgeelong.com/lineguide.php?line=geelong). railgeelong.com. . Retrieved 2008-03-28.

[3] "Rail Geelong - Geelong Line Events" (http://railgeelong.com/lineguide.php?line=geelong&section=events). railgeelong.com. . Retrieved 2008-03-28.

[4] "The Geelong Story" (http://www.historyvictoria.org.au/pdf/Geelong.pdf). History Victoria (http://www.historyvictoria.org.au). . Retrieved 2006-06-07.

External links

- Rail Geelong: Line History (http://railgeelong.com/lineguide.php?line=geelong)

V/Line VLocity train near Little River

N class locomotive hauled train running near Lara

V/Line train crossing the Breakwater Bridge

Colac station goods shed and platform

Sydney–Melbourne_rail_corridor

The **Sydney–Melbourne rail corridor** is an approximately 960-kilometre (600 mi) standard gauge railway corridor that runs between Melbourne and Sydney,[1] the two largest cities in Australia. Freight and passenger services operate along the route, such as the CountryLink XPT passenger service. The XPT offers a day or night service in each direction.

The railway corridor consists of NSW's Main Southern railway line from Sydney's Central Station to Albury Station – 646 kilometres (401 mi) – together with Victoria's North East railway line – 316 kilometres (196 mi) – from Albury Station to Melbourne's Southern Cross station (via Tottenham).

History

In 1883 the Victorian Railways broad gauge line met the New South Wales Government Railways standard gauge line at Albury, New South Wales at a break-of-gauge.[2] The two tracks were separated by a long island platform.

It was not until 1962 with the opening of the North East standard gauge line in Victoria that through services were possible between Melbourne and Sydney.[2]

The corridor was once home to intercapital passenger services such as the *Spirit of Progress*, *Southern Aurora* and *Intercapital Daylight*.

Current passenger services

The CountryLink XPT service runs two return trips each day between Melbourne and Sydney, making scheduled stops at Benalla, Wangaratta, Albury, Wagga Wagga, Junee, Cootamundra, Yass Junction, Goulburn, Moss Vale, Campbelltown and Strathfield with optional stops at Culcairn, Henty, The Rock, Harden and Gunning. Two additional stops in Victoria should commence in 2011. These are Broadmeadows and Seymour.

The CountryLink XPT provides train services between Sydney and Melbourne each day.

The day service from Melbourne to Sydney is the slowest scheduled service of the 4 XPT services at about 11.5 hours. The other three trips are scheduled to take about 11 hours. The discrepancy is due to the northbound daylight service being scheduled to wait in a loop south of Junee for the southbound service to cross.

Shortcomings

While the standard gauge line is approximately 960 kilometres (600 mi) long, the shortest distance available by road between Melbourne and Sydney is approximately 870 kilometres (540 mi) as the Hume Highway by-passes many of the towns on the rail route. Further evidence of the disparity in investment between rail and road can be seen in the fact that the Hume Highway is not far from having two lanes in each direction and being fully grade separated, whereas the rail is still single track in some places, containing over 50 level crossings in Victoria alone.

Sections of the rail line in NSW are signalled for a top speed of 65 kilometres per hour (40 mph) due to curves with small radii (some are 280 metres (920 ft)) and track with steep gradients (some are 1:38 adjusted for curvature).[3]

Freight trains struggle to maintain this speed in places. The number of curves along the entire route have been calculated to be equivalent to rotating the train by 72 circles (36 circles to the right and 36 circles to the left),[4] however due to the Bethungra spiral on the Sydney-bound track, there is an extra circle to the right for trains traveling from Melbourne to Sydney.

Improvement works

The corridor has long been double track from Sydney to Junee while the remainder of the route has been single track with a number of short loops.

Currently 200 kilometres (120 mi) or so of broad gauge track between Seymour and Wodonga is being standardised to form extra double track. The single track splits north of Seymour. There is a double crossover 3 kilometres (1.9 mi) further north, then another double crossover before Benalla, another before Wangaratta and another at Wodonga West. A 5-kilometre (3.1 mi) double track northern bypass of Wodonga has been constructed. This becomes single track just south of the Murray River and Albury.

A number of passing loops of about 7 kilometres (4.3 mi) length each have been added between Junee and Albury and between Seymour and Craigieburn. Also, some broad gauge track between Jacana and Albion has been made dual gauge to facilitate a new passing loop.

The four new passing lanes in Victoria include: i) Albion-Jacana ii) Donnybrook iii) Kilmore East iv) Tallarook.

The five new passing lanes in NSW include: i) Gerogery ii) Culcairn iii) Yerong Creek iv) Uranquinty v) Bomen.

The above changes, along with full concrete sleepering of the line and many signalling upgrades should be completed by late 2011.

More recently old 47kg rail is being replaced between Melbourne and Albury with new 60kg rail. This is due to be completed by November 2011.

Large sections of track have had mudholes appear since the concrete resleepering. This is associated with rough riding. Speed restrictions have been placed on affected track, and drainage works and fresh ballast are being applied to repair this.

As passenger trains get priority over freight trains, a freight-only track known as the Southern Sydney Freight Line is being added in Sydney in order to allow freight trains to travel in and out of the freight terminals during passenger peak times. This work has been delayed due to necessary yet unplanned service relocation along the line.

Some works have also been made in Melbourne to improve the flow of freight trains into the port.

A rail overpass has been constructed which connects the Standard Gauge line near Sunshine with the Standard Gauge line near Brooklyn. This allows trains to travel between the North-South (Brisbane-Sydney-Melbourne) and East-West (Melbourne-Adelaide) rail corridors without having to change direction at Tottenham. This additional track completes a Standard Gauge Sunshine/Brooklyn/Tottenham rail triangle. This track was commissioned mid 2010.

Other proposals

Over the years a number of deviations have been proposed for the track between Junee and Sydney, including between Glenlee and Aylmerton (known as the Wentworth Deviation), Werai and Penrose, Goulburn and Yass (Centennial Deviation), Bowning and Frampton including a bypass of Cootamundra (Hoare Deviation), and Frampton and Bethungra (removal of the Bethungra Spiral).[5] The proposals would replace 260 kilometres (160 mi) of winding track with 200 kilometres (120 mi) of straighter, higher-speed track, saving travel time, fuel, brake wear and track maintenance. However the Australian Rail Track Corporation have only documented plans for a handful of minor deviations to be completed by 2014.[6]

Over the past 30 years, there have been numerous proposals for a high-speed rail line from Sydney to Melbourne via Canberra, which would allow transit between the two cities in as little as 3 hours. With Sydney-Melbourne being one of the world's busiest air corridors, such a proposal may be revisited as Sydney airport reaches capacity, or if the cost of oil were to escalate.

Less ambitious proposals have included a minor 9.2-kilometre (5.7 mi) Jindalee Deviation mentioned in a 2006 Ernst and Young Report. Naturally a slow evolution consisting of many short deviations which can provide benefits sooner will not be equivalent to a few large deviations which could provide bigger bypasses and greater benefit. However more ambitious proposals come with greater risk of projects being delayed or cancelled.

References

[1] "ARTC - Investment Strategies - Project Delivery" (http://www.artc.com.au/Content.aspx?p=158#sia). www.artc.com.au. . Retrieved 2008-03-29.
[2] "VR History" (http://www.victorianrailways.net/vr history/history.html). www.victorianrailways.net. . Retrieved 2008-03-29.
[3] "NSW Curve & Gradient Diagrams: Section 2 - South" (http://extranet.artc.com.au/docs/engineering/infra_config/cd_diagrams/nsw/section_2_south/section_2_south_S00.pdf) (PDF). . Retrieved 2010-11-01.
[4] "Parliament of New South Wales: Road Transport (General) Amendment (Heavy Vehicle User Charges) Bill 2007" (http://www.parliament.nsw.gov.au/prod/PARLMENT/hansArt.nsf/V3Key/LC20071204040). . Retrieved 2008-05-08.
[5] "Interstate Rail Network Study, Improvements in Superfreighter performance: Sydney - Brisbane and Melbourne - Sydney Corridors, 2001" (http://www.artc.com.au/library/Maunsell_final_report.pdf) (PDF). . Retrieved 2009-05-29.
[6] "2008-2024 Interstate and Hunter Valley Rail Infrastructure Strategy - Executive Summary" (http://www.artc.com.au/library/RIS_2.1 - IA EXEC.pdf) (PDF). . Retrieved 2009-06-06.

Shepparton_railway_line

Shepparton railway line, Victoria	
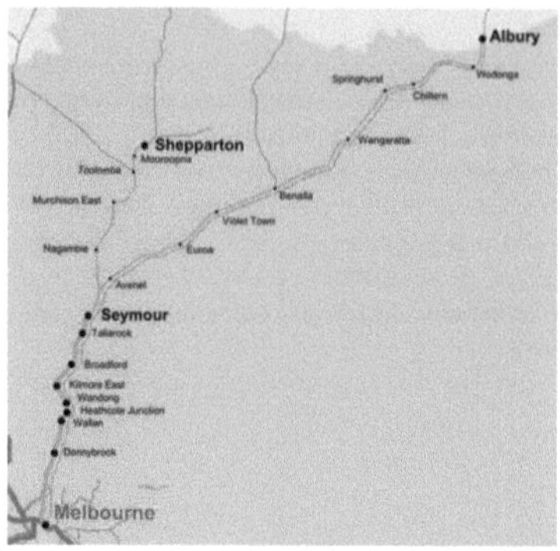	
Line details	
Opened	1860
Length	184.84 km to Shepparton 251.4 km to Tocumwal
Stations	12
Tracks	Double track to Seymour, Single track with passing loops beyond
Used by	V/Line and Pacific National
Service pattern	InterCity Limited stops between Melbourne and Seymour, stopping all stations beyond
Rolling stock	Locomotive hauled N type carriages
Connections	Seymour Line Albury-Wodonga Line
Rail transport in Victoria	
List of stations **Closed railways**	

The **Shepparton line** (also known as the Tocumwal line) is a 1600 mm (5 ft 3 in) railway line in Victoria, Australia. It branches off the Albury-Wodonga railway line at Seymour and has four passenger stations to Shepparton.

Line guide

The Shepparton line is served by daily V/Line services from Southern Cross Station in Melbourne. V/Line Seymour line services also operate along the inner section of the line, and is a branch line from the North East railway.

Bold stations are termini, *italic* stations are staffed at least part time (this has been confirmed).

History

The Melbourne and Essendon Railway Company opened the first section of the Shepparton railway line from North Melbourne to Essendon in 1860. Following its take over by the Victorian Government in 1867, the line was extended to Tallarook and Mangalore in 1872. This line was completed to Albury in 1883 (see Albury-Wodonga line).

A line was built from Mangalore to Toolamba and Shepparton in 1880 and extended to Numurkah in 1881, Strathmerton in 1905 and connecting with the New South Wales Government Railways at Tocumwal at a break-of-gauge in 1908.

Passenger services to Tocumwal ended on 8 November 1975 with the last train operated by T class diesel locomotive T324 and passenger carriages 3AS - 31BE - 2AE - 22CE. Before this time the Strathmerton - Cobram section of the line was operated as the 'branch line' with a 102hp Walker railmotor connecting with the main line train. A bus service was then introduced for the Tocumwal branch, connecting with the Cobram service. By 1977/78 the service between Cobram and Tocumwal was being operated by a VicRail owned station wagon driven by the Cobram station master.[1]

The last passenger service from Toolamba station to Echuca ran on 2 March 1981 with Y class diesel locomotive Y161, an ABE carriage and a C van. This consist had only been introduced a few months prior, with a DERM usually being rostered.[1] Toolamba finally closed as a station on 20 December 1987.[2]

The passenger service from Numurkah north to Cobram was withdrawn on 24 April 1981 but was restored on 14 August 1983.[1] The service from Melbourne to Shepparton then to Cobram was again withdrawn on 21 August 1993, with Hoys Roadlines taking control of the train as far as Shepparton from 22 August,[1] hiring locomotives, carriages and train crews from V/Line. This arrangement remained until 2004.[3]

There have been calls for the line to be standardised as part of the proposed Melbourne to Brisbane 'Inland Railway'. In April 2008 it was announced that the Shepparton – Tocumwal section of the line would also be upgraded, as part of the Victorian core grain network in a A$23.7 million package with 6 other lines.[4]

Branch lines

A branch line was subsequently built from Murchison East to Rushworth and Colbinabbin, with a branch from Rushworth to Stanhope and Girgarre. A line was completed between Toolamba and Echuca in 1887.

A short branch line was opened from Shepparton to Dookie in 1888. A private tramway was opened between Dookie and Katamatite in 1890, and was taken over by Victorian Railways in 1896.[5] Another short branch line was opened from Strathmerton to Cobram in 1888.

References

[1] Chris Banger (March 1997). "Rail Passenger Service Withdrawals Since 1960". *Newsrail* (Australian Railway Historical Society (Victorian Division)): pages 77–82.
[2] Bob Whitehead (2008). *Tocumwal Railway Centenary*.
[3] "Track Record: 19 April to June 2004" (http://www.doi.vic.gov.au/doi/doielect.nsf/2a6bd98dee287482ca256915001cff0c/a3da3b97a6cc8a96ca256f0b00112ea9/$FILE/trkrcd19.pdf). www.doi.vic.gov.au. . Retrieved 2008-07-13.
[4] "$43m to upgrade rail freight lines" (http://business.theage.com.au/43m-to-upgrade-rail-freight-lines/20080421-27nl.html). *The Age*. . Retrieved 2008-04-27.
[5] "Katamatite" (http://web.archive.org/web/20060821093130/http://www.arts.monash.edu.au/ncas/multimedia/gazetteer/list/katamatite.html). Australian Places (http://www.arts.monash.edu.au/ncas/multimedia/gazetteer). Archived from the original (http://www.arts.monash.edu.au/ncas/multimedia/gazetteer/list/katamatite.html) on 2006-08-21. . Retrieved 2006-06-07.

External links

- http://www.vline.com.au (http://www.vline.com.au)
- Official map (http://www.vline.com.au/maps/maps/northeast.html)
- Statistics and detailed schematic map (http://www.vicsig.net/index.php?page=infrastructure&section=lineguide&line=Tocumwal) at the vicsig (http://www.vicsig.net/) enthusiast website

Melbourne–Adelaide_railway

Melbourne-Adelaide railway	
Locale	Victoria and South Australia
Predecessor	Victorian Railways and South Australian Railways
Track gauge	1435 mm (4 ft 8 $^1/_2$ in) (standard gauge), converted from1600 mm (5 ft 3 in)

The **Melbourne–Adelaide railway** is a standard gauge railway corridor that runs between the cities of Melbourne, Victoria and Adelaide, South Australia.[1] Most of the current traffic is freight, though the only named and perhaps best known regular train is the thrice-weekly passenger service *The Overland* operated by Great Southern Railways

History

In the second half of the nineteenth century the Victorian Railways and South Australian Railways broad gauge networks were extended. The South Australian main line, the Adelaide-Wolseley railway was connected to the Victorian system at Serviceton in 1887. This was the first single gauge inter-colonial link in Australia.[2]

Conversion to standard gauge and rerouting

In 1995 the line was converted to standard gauge under the One Nation program and rerouted between Melbourne and Ararat in Victoria to run on the line via North Shore and Cressy.[3]

Track and gauge

The line is single track the entire route with the exception of a short dual gauge section near Melbourne and a number of 1500-1600 m crossing loops every 15-45 km.

References

[1] "AusLink Network Corridors" (http://web.archive.org/web/20080719153251/http://www.auslink.gov.au/whatis/network/corridors/AusLink_Corridors.aspx). www.auslink.gov.au. Archived from the original (http://www.auslink.gov.au/whatis/network/corridors/AusLink_Corridors.aspx) on 2008-07-19. . Retrieved 2008-03-29.

[2] "ARHS Railway Museum: History 1839 - 1900" (http://www.railwaymuseum.org.au/history.html). www.railwaymuseum.org.au. . Retrieved 2008-03-29.

[3] "ARHS Railway Museum: History 1950 - now" (http://www.railwaymuseum.org.au/history3.html). www.railwaymuseum.org.au. . Retrieved 2008-03-29.

Western_standard_gauge_line

Western standard gauge railway line, Victoria	
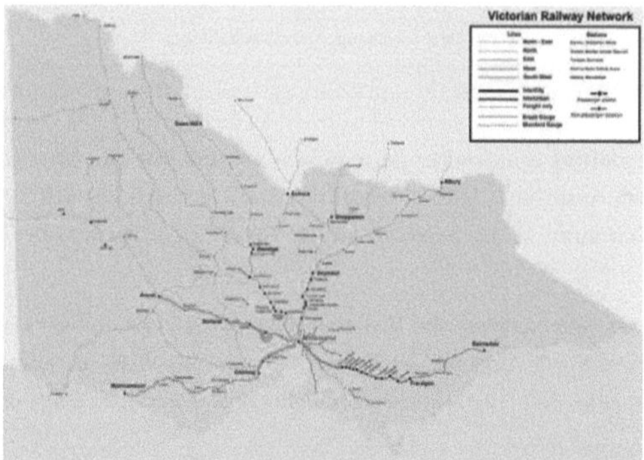	
Line details	
Commenced	1889
Completed	1995 (in current form)
Tracks	Single track, crossing loops
Used by	Freight and *The Overland*
Connections	Portland, Hopetoun and Yaapeet lines
Former connections	Bolangum and Carpolac lines
Rail transport in Victoria	
List of stations **Closed railways**	

The **Western standard gauge** is a railway line in western Victoria, Australia. Opened in 1995, it forms part of the Melbourne-Adelaide railway and serves as the principal interstate rail link between Victorian and the western states. The line replaced a number of former broad gauge routes which were gauge converted, and today sees both intrastate and interstate freight traffic, as well as the thrice weekly (in each direction) *The Overland* passenger service. Major towns on the route include Geelong, Ararat, Horsham and Dimboola.

History

The first intercapital link between Melbourne and South Australia was completed in 1887 when the Victorian Railways line was extended to Serviceton on the state border.[1] Known as the Serviceton line, it passed through Geelong, Ballarat, Ararat, Stawell, Horsham and Dimboola, on the way west. It was not until 1889 that a direct Melbourne - Ballarat link was opened.[2]

In the 1970s most interstate lines in Australia began to be converted to standard gauge. By the 1990s Adelaide to Melbourne was the only interstate link not converted, and so various proposals were made for gauge conversion. Two main options were put forward:

- via Ballarat: A new track or dual gauge to Ballarat, then conversion of the line west.

- via Geelong: A new track to Geelong, dual gauge to Gheringhap, then conversion of the line via Cressy, and then conversion of the line from Ararat.

Various reasons for given for and against both routes. The Geelong was longer, but avoided the steep grades on the line though Ballarat. Businesses and industry in both cities also wanted to be on the main interstate link. In the end it was decided to build the route via Geelong, and work was completed in 1995 with funding from the Federal Government One Nation program. Along with the gauge conversion of the interstate line, the Portland, Yaapeet and Hopetoun lines were also converted, and a dual gauge link provided to Maryborough to permit grain from the north-west to reach the port at Portland.

Infrastructure

The line is standard gauge, except for dual gauge on the Newport - Sunshine freight line, and where it follows the Geelong - Ballarat line. It is single track throughout, with numerous 1500 metre long crossing loops.[3] The majority of the line uses Centralised Traffic Control to direct trains, except for the Gheringhap to Maroona section which uses the Section Authority System,[4] a Victoria-only radio token based safeworking system developed in the 1980s. The line is owned by VicTrack and since 1997 has been managed by the Australian Rail Track Corporation.[5]

Services

The main traffic on the line is Melbourne - Adelaide interstate freight, with trains operated by Australian Railroad Group, Pacific National and Specialised Container Transport. Grain services are also operated on an irregular basis by Pacific National, Genesee and Wyoming Australia and El Zorro.[6]

A local container freight service also operates from Melbourne to Horsham by QRNational for Wimmera Container Lines.[7] Until July 2008 it was operated by Pacific National, who cancelled it in April then gave it a three month reprieve.[8] [9] The only passenger service is *The Overland* thrice weekly.

Route

The line leaves the North East standard gauge line at Tottenham, then runs south via the dual gauged Newport - Sunshine freight line. From Newport a separate line is provided to the west of the Geelong line as far as North Geelong Junction, where the line joins the dual gauged Geelong - Ballarat north as far as Gheringhap Loop. From here line follows the gauge converted Gheringhap - Maroona line westward until it meets the Portland line, and heads north to Ararat where it rejoins the former main line.

Blue is standard gauge, red is broad gauge. **Bold** are active passenger platforms.

References

[1] "ARHS Railway Museum: History 1839 - 1900" (http://www.railwaymuseum.org.au/history.html). www.railwaymuseum.org.au. . Retrieved 2008-03-27.

[2] "VR History" (http://www.victorianrailways.net/vr history/history.html). www.victorianrailways.net. . Retrieved 2008-03-27.

[3] "Appendix II Interstate Network Overview" (http://www.artc.com.au/library/TA02 Appendix II Issue 2.2.pdf). *ARTC - Access Seeker Network Configuration and Description*. www.artc.com.au. . Retrieved 2008-03-28.

[4] "VICSIG: Western SG Line" (http://vicsig.net/index.php?page=infrastructure&line=Western SG). vicsig.net. . Retrieved 2008-03-28.

[5] "ARHS Railway Museum: History 1950 - now" (http://www.railwaymuseum.org.au/history3.html). www.railwaymuseum.org.au. . Retrieved 2008-03-28.

[6] "Operators, Trains and Train Numbers" (http://members.wimmera.com.au/wangara/west/trains.html). members.wimmera.com.au. . Retrieved 2008-07-30.

[7] "Rail to roll on" (http://wimmera.yourguide.com.au/news/local/news/general/rail-to-roll-on/1229118.aspx). *Wimmera Mail Times*. yourguide.com.au. 2008-07-28. . Retrieved 2008-07-30.

[8] "Rail freight woes spark crisis of confidence" (http://www.abc.net.au/news/stories/2008/04/21/2222687.htm). *ABC News*. www.abc.net.au. . Retrieved 2008-07-16.

[9] "Three month extension for Wimmera rail freight services" (http://www.abc.net.au/news/stories/2008/04/24/2226019.htm). *ABC News*. www.abc.net.au. April 24, 2008. . Retrieved 2008-07-16.

Article Sources and Contributors

Outtrim_railway_line *Source*: http://en.wikipedia.org/w/index.php?title=Outtrim_railway_line *Contributors*: Amalas, Antonia palmieri, Axver, Billingd, Coyets, Crusoe8181, Dan027, Infrangible, Iridescent, John of Reading, Likeslife, Nomadtales, Peter Campbell, Peter Horn, Platte, Rich Farmbrough, Sameboat, Supt. of Printing, Weakopedia, 1 anonymous edits

Outtrim,_Victoria *Source*: http://en.wikipedia.org/w/index.php?title=Outtrim%2C_Victoria *Contributors*: Billingd, Crusoe8181, Drpickem, Peter Campbell

South_Gippsland_railway_line *Source*: http://en.wikipedia.org/w/index.php?title=South_Gippsland_railway_line *Contributors*: Angela26, Bleakcomb, Coyets, Crusoe8181, Dan027, Darcycammo, DoubleBlue, Grahamec, Lakeyboy, LilHelpa, Mboverload, Merlion444, Michael Bednarek, Nasnema, Neddy53, Nomadtales, Peter Horn, PigFlu Oink, Rich Farmbrough, Rjwilmsi, Signalhead, Slabba, Tanarrifujitsu, Tds247, Thechristoff, Thiseye, Tom Morris, WereSpielChequers, WikHead, WilliamGrant, Wongm, Zzrbiker, 100 anonymous edits

Pakenham_railway_line *Source*: http://en.wikipedia.org/w/index.php?title=Pakenham_railway_line *Contributors*: 712M, Afromcbenny, Australian Jezza, AxSkov, B.d.mills, Colonies Chris, Coyets, Crusoe8181, Crzrussian, Ewawer, Fcfhkmelb, Grimhim, Koavf, Lahroo, Mr.schnoobley, Ozzmosis, Peter Horn, Philip J. Rayment, R'n'B, Rigadoun, Rom rulz424, Sameboat, Sb617, Sheepunderscore, Skapur, Somebody in the WWW, Supt. of Printing, TPK, Thewinchester, Thryduulf, Trainz10, Wongm, Woohookitty, Wwoods, 45 anonymous edits

Rail_transport *Source*: http://en.wikipedia.org/w/index.php?title=Rail_transport *Contributors*: *drew, -Majestic-, 1122334455, 12.234.49.xxx, 16@r, A bit iffy, Aaaallleex, Abby, Acela2038, Acprail, Acroterion, AdamW, Adashiel, Ae-a, Aesopos, Ahoerstemeier, Ahuskay, Aitias, Alansohn, Alco83, Aleator, AlexPlante, Alexdan loghin, Alextrevellan 006, Allreet, AllyUnion, Altenmann, Alucard (Dr.), Amakuru, AmyzzXX, Andreas Kaganov, Andrewlim1, Andrewpmk, Andy Dingley, Anlace, Anne McDermott, Antandrus, Anthony Appleyard, Armando, ArnoldReinhold, Arsenikk, Arthur Ellis, Aseemsjohri, Ashishbijwe24, Aude, Autonerd, Avenue X at Cicero, AvicAWB, AxelBoldt, Ayushrocks6, BD2412, BRG, Backifran, Bambuway, Basketball1776, Bastin, Bathrobe, Bebestbe, Beland, Berendale2, Bermicourt, Besselfunctions, Beyond silence, Bhadani, Bharath93, Bhtpbank, Bigdumbdinosaur, Biggus wickus, Biker Biker, Bikramshergill, Bill37212, Birdhurst, Biscuittin, Bkell, Black Kite, Blehfu, Blimpguy, Bluerasberry, Bob Castle, Bobo192, Bogdangiusca, Borgx, Brambo, Brhaspati, Brian Pearson, Brian evans, BruceDLimber, Bryan Derksen, Btornado, Bulleid Pacific, Burgher, Camboxer, Can't sleep, clown will eat me, CanadianLinuxUser, Canterberry, Casey56, Caseyjonz, Cazort, Cecropia, Celardore, Central Data Bank, Cessator, Chaleyer61, ChanceButler, Charlottesometimes, Cheekdog, Chevin, Chris55, Christian List, Chun-hian, Civil Engineer III, Claude girardin, Cluth, Cmapm, Cmichael, Colonies Chris, Complete fanatic, Conversion script, Coolhawks88, Coosbane, Corpx, Corti, Courcelles, CrossHouses, CryptoDerk, Cwolfsheep, Cynical, DAJF, DJ Sturm, Damirgraffiti, Dancter, Daniel C, Dark Shikari, DarkFireTaker, Das48, Daveswagon, David.Monniaux, DavidArthur, Ddstretch, Dealerofsalvation, Dearmstrong, Deb, Dellarb, Delldot, DennisJOBrien@yahoo.com, Dethme0w, Devin.phillips, DianeSelvy, DigbyDalton, Dnalrom123, Dnfenner, Doc glasgow, Download, Dp67, Dragomiloff, Dratman, Dsteckelberg, Duncharris, Dysprosia, ESkog, EarthFurst, EdJogg, Edgar181, El C, ElectraFlarefire, Emdx, Emperor Genius, Emsox, Engi08, Epbr123, Epolk, Equinox137, Eric-Wester, Exe, Expertz123, Ezhiukas 7, FT2, Faradayplank, Farjiadmi, Ferrierd, Feydey, Finavon, Finngall, Fixerofnatasha, FlavrSavr, Flux.books, Foodman, Fortdj33, Frankenpuppy, Fred Bauder, Fredericsalve, Frixa, Fudoreaper, Furrykef, Fyrael, G-Man, Gandydancer, Gary D, Gbtojun, Gdo01, Geoff NoNick, Gerfriedc, Geschichte, Ghaly, Glane23, Globalsolidarity, Golbez, Grafen, Graham87, Grim23, Gsaup, Gtstricky, Gun Powder Ma, Guybrarian, Gwernol, Gökhan, Hairy Dude, Harjasusi, Haroonn1, Harpblaster, Harperbruce, Hclincha, HenryLi, Hephaestos, Hmrox, Hoo man, Hu, Hugh Manatee, Hulagutten, Hydriotaphia, Ianp1950, Ichol, Ishlan, J.delanoy, JCam, JNW, JRR Trollkien, JYolkowski, JackLumber, Jacob.jose, Jallason, Jameswilson, Japman6, Jarle fagerheim, Jax184, Jclemens, Jcuk, Jdcook, Jdforrester, Jeepday, Jeffakolb, Jeremy5561, Jevon, Jhonchris1981, Jim, Jimp, Jinty182, Jmorgan, JoeK, Joethespaz13, Joey Roe, John, JohnOwens, Johnasher, Jojit fb, Jonathan.s.kt, JorgeGG, Joshua, Jovianeye, Jpgordon, Jpk, Jrleighton, Jstreutker, Juggleandhope, JuliusAugustus, Jusjih, Juxo, KF, KVDP, Kahuzi, Kaibabsquirrel, Kalemika, Kalixiri, Katsp8, Kazvorpal, Kdar, Keeper76, Keidansky, Keithmall, Kelisi, Kelly Martin, Keotaman, Ketanpujare, Kingpin13, Kkmd, Krath, Krigjsman, Ksyrie, Kumioko, Kungfuadam, Kuru, L blue l, La Pianista, Lafleure355, Lars T., LeaveSleaves, Leefletcher, LeoO3, Leonard G., Leptictidium, Levineps, Lightmouse, Lindahq, Lockesdonkey, Longhair, Lord Bob, Loren36, LorenzoB, Lost on belmont, Lost tourist, Lot29, Lovemehateme123, Luna Santin, MNboy, MPD01605, Madcoverboy, Madhero88, MakeChooChooGoNow, Mamawrites, Man vyi, Mangoe, Manop, Maranjc, MarsRover, Marshman, Matt von Furrie, Mattisse, MauriceJFox3, Maury Markowitz, Mav, Mayooranathan, Mdd4696, Megaman en m, Mervyn, Mgaved, Michael Hardy, MichaelBillington, Michaelduly, Mike Dill, Mikeasolomon, Mindspillage, Minesweeper, Mitul0520, Mo0, Moebiusuibeom-en, Moncrief, Monedula, Morven, Moyogo, Murray Langton, N2e, NE2, Naddy, NameThatWorks, Nanana123456789, Nat Krause, Nerfer, NicholasJones, Nick Number, Nicolae Coman, Nk, Noexit, Noisy, Northumbrian, NotWith, Nricardo, Nunh-huh, Oanabay04, Oceanhahn, Olahus, OlavN, Olivier, Omnedon, Ortolan88, PFHLai, PNRStreamliner, PReinie, Pak bahadur, Patoune08, Patrick, Paul Matthews, Paul W, PaulDrye, Perdelsky, Peter Horn, PeterEastern, Peterkingiron, Pgk, Phlebas, Piano non troppo, Pickle UK, Pietrow, Pilgaard, Pilotguy, Pinethicket, Pion, Piotrus, Playclever, Plutochaun, Pne, Pointer1, Pressforaction, ProhibitOnions, Prolog, Propaniac, Psublue23, Qertis, QuantumEngineer, Quintote, R'n'B, RFBailey, RTucker, Rafikk, Randall uob, Redrose64, Regregex, Rememberway, RevRagnarok, Reward, Rich Farmbrough, Rich4560, Rickyrab, Rigadoun, Rjensen, Rjmccall, Rjstott, Rjwilmsi, Rmhermen, RobHarding, Robert Merkel, Rods42, Rojomoke, Roke, Rrburke, Rror, Rsrikanth05, Runewiki777, S.K., SEWilco, SNIyer12, SPUI, ST47, Salamurai, Sam Hocevar, Sameer rana100, Samuel 69105, Samuell, Sandman30s, Sarah Waggoner, Sarah tonen, Satori Son, Sc147, ScAvenger lv, Scoot-Overload, Scoutersig, Seaphoto, SebbeSwe, Sentorment, Sergio.solar, Sgfaig, Sharadbob, SheepNotGoats, Sheldwich14, Signalhead, SilkTork, Silverjonny, SimonP, Skapur, Slambo, Smalljim, Smerlinare, Smurrayinchester, Snottily, Snowmanradio, Socipoet, Soliloquial, Sonett72, Sp33dyphil, Spagf5, Spangineer, Spearhead, Spellchecker, SqueakBox, Squids and Chips, SriMesh, Srtig, Starbois, Steam-loco, Stefan2, Steinberger, Stephen C. Carlson, Stragermont, Struway2, Stumpytrain, Sulzer55, Supt. of Printing, Suri 100, Svenman, Svetovid, SwalleyD, Syd1435, Symane, TAS, THEN WHO WAS PHONE?, Tabletop, Tagishsimon, TaintedMustard, Takamaxa, Taksim25, Tangent747, Taoster, Tarquin, Tassedethe, Tatakubuntu, Tcr25, Tedernst, Tellyaddict, Template namespace initialisation script, TerryElliott, Textorus, Tgv8925, The Arbiter, The Wild Falcon, The undertow, TheBourtreehillian, TheEgyptian, TheListUpdater, Theda, Thephotoplayer, ThomasAndrewNimmo, Tide rolls, Tim PF, Timc, Tmorgan147, Toiyabe, Tony1, TonyTheTiger, Train2104, Trekphiler, Ttsalo, Tudorminstrel, Ulric1313, Undead1, Vapmachado, Vcwizard, Vegas949, Villager57, Voyager, Vuo, WOSlinker, Wahiba, Wavelength, Wdfarmer, Webnetprof, Wereon, WiJG?, WikHead, Wiki Raja, Wikipelli, William Grimes, Wolfkeeper, Wongm, Woodenships, Woohookitty, Wspencer11, Wuhwuzdat, Yelyos, Yossarian, Yvwv, Zhou Yu, Zoney, Zzuuzz, ŠJů, Јованвб, 825 anonymous edits

Victoria_(Australia) *Source*: http://en.wikipedia.org/w/index.php?title=Victoria_%28Australia%29 *Contributors*: 0, 142.177.55.xxx, 144.132.70.xxx, 1manfern, 1subwoofer, 203.37.81.xxx, 3DBrendzz, 5.4, 52 Pickup, 61.9.128.xxx, 7zara7, Aaroncrick, Abhijitsathe, Acad Ronin, AcrophobicEagle, Adi4094, Adriankitch, Adz, Al Silonov, Alansohn, Albert Isaacs, Aldie, Alex3895, AlexiusHoratius, Aliasd, AmishThrasher, Amorim Parga, Anandks007, Andre Engels, Andrija987, Andyman14, Angusmclellan, AnnaVida12345, Anonymoues, Ansett, Antandrus, Antipodean Contributor, ApprenticeFan, Artemus Jones, Auriong, Aussie.power, AussieLegend, Australiana, Avala, Averagejefpet, Avoided, AxSkov, B.d.mills, Backslash Forwardslash, Bahudhara, Barneca, Barrylb, BazookaJoe, Bduke, Beancadidor, Bearcat, Bearo, Ben Ben, BennyBJames, Biatch, Bidgee, Blameytherinosaur, Bleakcomb, Blueracer6, Bobblewik, Bobo192, Bongomanrae, Bongwarrior, Brambleclawx, Breno, Brian0918, Brianga, Brighterorange, Buaidh, Bugsplatterspickspick, Burgundavia, Burnwelk, CJ, CTVDigital1, Caiaffa, Calsicol, Camerong, CanadianLinuxUser, Canglesea, Canley, Cassowary, Chamdarae, Chanheigeorge, Chicken7, Chiken23, Chris the speller, Chrisleigh, Chrism, Christopher Parham, Chuq, Civil Engineer III, Cjmhall, Cjmnyc, Ckatz, Clarkk, Cmathio, Coffee, Comes.amanuensis, CommonsDelinker, Conversion script, Coredesat, Crazycrazyduck, Crusoe8181, Curieux, Curps, Cyanolinguophile, DTGHYUKLPOQWMNB, Darth Panda, Darwinek, David Gerard, David Kernow, Davidzuccaro, Dcandeto, Dcashm, Deanos, Decltype, Delldot, Dementured, Derek Parnell, Dethroned Buoy, Discospinster, Dmondark, Doggy10113649, Doug, Dragases, Drenver, Dubc0724, Duckie7, EGetzler, Easel3, Egmontaz, Elekhh, Epbr123, Epistemos, Eregli bob, Erianna, Ewawer, Extra999, Falcadore, Felix Dance, FetchcommsAWB, Fider, Figaro, Fikri, FlavrSavr, Flyguy649, Fred31, GSTQ, Garth M, Gaudoine, Gdo01, Geekyloveread, Genovue, GenuineMongol, Geoffmark, Gigemag76, Givern, Glst2, Gobillino, Gogo Dodo, Goldfishbutt, Good Olfactory, Grafen, Graham87, Grahamec, Greentubing, Greglocock, Griffinofwales, Guanaco, Gurch, Gwernol, Hachiko, Hans Dunkelberg, Harley3904, Harmil, Hayden120, HelloMojo, Hesperian, Heymiki87, HiLo48, HillsOfEndeavour, HorseloverFat, Hsan22, Hunarian, Ian Dunster, Ianblair23, Icairns, Icseaturtles, Illawong, Imroy, Island, Istabraq, Izimini, J.delanoy, JPD, Jackelfive, James000, James42.5, Jameslamb, Jauhienij, Jayand, Jc4p, Jj137, Jklin, John Vandenberg, John254, JohnnoShadbolt, Johnny C. Morse, Johnpellegrino, Joseph Solis in Australia, JoshuaZ, Just Another Dan, Jutherine, Jwoodger, KAPITALIST88, KaragouniS, Karen Johnson, Kendo207, Kesuari, Kewp, Kilhan, Kingpin13, Knobbly, Konstable, Kubigula, Kyle Andrew Brown, Lee M, Lightmouse, Lil jeddy, LilyU1, Lisiate, Littlest Plum, Liveinvictoria, Lloyd borrett, LocalRes, Lollol2134, Loltingz, Longhair, Lou.weird, LovesMacs, Luuva, M.O.X, MER-C, MJCdetroit, MacTire02, Maias, Mandarax, Marco polo, Marek69, Mark, Marmalade999, Martin23230, Matilda, Mattinbgn, Mattnotley2004, Maximus Rex, Maxis ftw, Mboverload, Meezaa., MegaSloth, MelbourneStar, Melburnian, Melchoir, Merbabu, Messenger88, Michael Johnson, Michaelisgun, Minesweeper, Mitch Ames, Mitsuhirato, Mkativerata, Mmcqueeney, Momo san, Moniquecharlotte, Moondyne, MoondyneAWB, Moorstreet, Mr Gronk, Muijz, Muzi, Mvjs, MyallR, N5iln, NHRHS2010, Nachoman-au, Narracan3824, Natronomonas, NatureA16, Nburden, NecroSlayer1490, Newone, Nick carson, Nick-D, Nikitagentle, Nilfanion, Nippymippy, Nix35, NorwegianBlue, NouvelleAuteur, Novakreo, Nricardo, Ohconfucius, Ontonno, Orderinchaos 2, Ozzieboy, Ozzmosis, PDH, Patar knight, Patrick, Pax:Vobiscum, Paxsimius, Pberrett, Peteboro, Peter I. Vardy, Pgk, Princess Maria, Purrum, Pursey, Quantumobserver, Qwertywop, Raider2044, Rainer, RandomAct, Randwicked, Ranveig, Raven4x4x, Rebecca, Recognizance, Recurring dreams, RedWolf, Regibox, Rehevkor, Reinthal, Rettetast, RexNL, Richard Cranston, Richde, Risker, Riumplus, Rlee0001, Rm w a vu, RobDe68, Robert Merkel, Rodolph, Roisterer, Roke, Rrius, Rulesfan, Rusty201, Ryulong, S in mtl, Sam Blacketer, Samdutton, Sardanaphalus, SatuSuro, Sb617, SchfiftyThree, Schwnj, Scottius11, Seaphoto, Seattle Skier, Secfan, Seqsea, Sfaiku, Sgall, Sha0000, ShardPhoenix, Shogun, Silversmith, Skyezx, Slawojarek, Smarkflea, Smoore 500, Sp33dyphil, SpookyMulder, Srippon, StAnselm, Stan Shebs, Starshadow, Stephen, Stephenb, Strum that guitar, Stupidmunani, Suffusion of Yellow, SuperHamster, Supt. of Printing, Swarve, Sweetcutiepie, Syrthiss, TPK, Tabletop, Tamasys, Tannin, Tawker, Tedp, Template namespace initialisation script, Tessaholmes, The Emirr, TheMaskedImprover, TheProject, TheReaderOfOz, Theleftorium, Theodolite, Theoneintraining, Thingg, ThomasPusch, Thomasgl, Thx adidas 1138, Tide rolls, Tiight, Till I Go Home, Tim Starling, Tirin, Tobias Conradi, Tony1, Trek011, Ulric1313, Unseen12345, Vcaaweb, Vegaswikian, Vicki Virago, Viriditas, WWGB, Wachholder0, Walton One, Wavelength, Wcrosbie, Weedwhacker128, Wheeler1346, WhisperToMe, Widders, WikHead, WikiTownsvillian, Wikivolunteer, WildBlue, Wongm, Wwoods, X!, Xionbox, Xiroth, Xtra, Xyzzyva, Yassie, YellowMonkey, Yerpo, Yes four, Ylee, Zacharybowen, Zamias, Zigger, Zoe, Zootalures, ΚΕΚΡΩΨ, 770 anonymous edits

Dandenong_railway_station *Source*: http://en.wikipedia.org/w/index.php?title=Dandenong_railway_station *Contributors*: 712M, Babomb, Ben Ben, Bluemoose, Breno, Coyets, Crusoe8181, Crzrussian, Deor, Elonka, Epistemos, Evan C, Farsouth, Grimhim, Grutness, Henry W. Schmitt, Huyie, Invincible, Jamesbehave, Kleopatra, Lakeyboy, Longhair, MelbourneStar, Ozzmosis, Pi lambda, Rjwilmsi, Rom rulz424, Sandgem Addict, Sb617, Signalhead, Somebody in the WWW, TPK, The Funeral Party, Thewinchester, Timorous, Ttomau, Wai Hong, Wongm, 35 anonymous edits

South_Gippsland *Source*: http://en.wikipedia.org/w/index.php?title=South_Gippsland *Contributors*: Biatch, Pengo, THEN WHO WAS PHONE?, 4 anonymous edits

Korumburra_railway_station *Source*: http://en.wikipedia.org/w/index.php?title=Korumburra_railway_station *Contributors*: AaronRichard, Amalas, Billingd, Crusoe8181, D6, Dan027, Grey Shadow, Rebecca, Thechristoff, Wongm, 30 anonymous edits

Korumburra,_Victoria *Source*: http://en.wikipedia.org/w/index.php?title=Korumburra%2C_Victoria *Contributors*: Aimeebrew, Angela, AussieLegend, Avicennasis, Berryscreek, Cmichael, Cnwb, Crusoe8181, Davey the pirate, EJF, Epistemos, Gaius Cornelius, Grahamec, Jevansen, John of Reading, Kmlse2348106, Lcarscad, Longhair, M499, Mattinbgn, Melburnian, Michael Johnson, MichaelBillington, Mindmatrix, Mlpearc, NeoJustin, Orange Suede Sofa, Pan Dan, Pinethicket, Purrum, Revolutionary penguin, Reynardo, Roger godfrey, Romana32, Sgt Pinback, Shanel, Slabba, Stuffnjunk, The Thing That Should Not Be, Tide rolls, 62 anonymous edits

Victorian_Railways *Source*: http://en.wikipedia.org/w/index.php?title=Victorian_Railways *Contributors*: 7, Antonia palmieri, Bidgee, Boleyn3, Clarkk, Dan027, Ewawer, Garyvines, Giggler99, Graeme Bartlett, Grahamec, Gunzel, Johnno128, Kuliwil, Lancybaby, Liamdavies, Longhair, Mattnotley2004, Michael Johnson, Mitchell price, Motacilla, Philip J. Rayment, Phoe, Rjwilmsi, SatuSuro, Supt. of Printing, Thadocta, Trainz10, WikiSandbox1, Wongm, Wuhwuzdat, Zzrbiker, 27 anonymous edits

Victorian_Heritage_Inventory *Source*: http://en.wikipedia.org/w/index.php?title=Victorian_Heritage_Inventory *Contributors*: Billingd, Crusoe8181, SatuSuro

List_of_closed_railway_stations_in_Victoria *Source*: http://en.wikipedia.org/w/index.php?title=List_of_closed_railway_stations_in_Victoria *Contributors*: Ae-a, Amalas, Bender235, Biatch, Crusoe8181, Dan027, Dddstone, Epistemos, Forthevline, Hugo999, Jason Recliner, Esq., Longhair, Mcauchi, Netoholic, Optimist on the run, Orderinchaos 2, Peter Horn, Rebecca, Rich Farmbrough, Rick Doodle, Somebody in the WWW, Tabletop, Thryduulf, Tomas86, Wavehunter, WikiSandbox1, Wongm, 39 anonymous edits

Jumbunna_railway_station *Source*: http://en.wikipedia.org/w/index.php?title=Jumbunna_railway_station *Contributors*: Amalas, Billingd, Crusoe8181, Dan027, Longhair, PKT, Rich Farmbrough

Transport_in_Australia *Source*: http://en.wikipedia.org/w/index.php?title=Transport_in_Australia *Contributors*: 1717, Adam.J.W.C., Adz, Andreasu, AndrewMcQ, Andypasto, Auroranorth, AussieLegend, AxSkov, Beland, Bermicourt, Biatch, Bobblehead, Borck103, Bryan Derksen, Camerong, CanadianLinuxUser, Cassowary, Chanlord, Chuq, Clarkk, Codeispoetry, Conversion script, Crusoe8181, Digitalparfait, Dl2000, Dr. Blofeld, Dtremenak, DuncanHill, Eastlaw, Ellmist, Eloquence, FNQ, Figaro, Fuhghettaboutit, Gaz, Gilliam, Gobeirne, Gobillino, Graham87, Grahamec, Gunzel, Gwernol, Harryboyles, Haus, Helpa26, Hesperian, Hossen27, ICarlz, Iridescent, J.delanoy, Jguk, Jpatokal, Kdurrance, Kerotan, Koyaanis Qatsi, KoyaanisQatsi, Kritterhound, MaXim, Mahahahaneapneap, Mardenpb1, Mark, Mark Ryan, Mattnotley2004, Michael Glass, Millere08, Mortein, Mvjs, Nachoman-au, Nadavspi, Neddy53, Neelix, Nick carson, Nomadtales, One-dimensional Tangent, Orderinchaos, Patrick, Paypay3, Philthecow, Piano01, Psychohug, Rebecca, Rich Farmbrough, Rjwilmsi, Robert Merkel, Salsb, Sannse, Sb617, ScottDavis, Seoidau, Shiftchange, Shinjiman, Somebody in the WWW, Speer320, Stemonitis, TPK, Tabletop, Tannin, Themeparkgc, Thingg, Thortful, Tim Starling, Tobias Hoevekamp, TruckCard, Wally Otto, Wangi, Warofdreams, Widetree45, Wiki ian, Wikitanvir, William Avery, Wongm, World8115, YSSYguy, Zigger, Zoicon5, Zondor, 103 anonymous edits

Outtrim_North_railway_station *Source*: http://en.wikipedia.org/w/index.php?title=Outtrim_North_railway_station *Contributors*: Amalas, Crusoe8181, Dan027, Longhair, Rich Farmbrough

Bendigo_railway_line *Source*: http://en.wikipedia.org/w/index.php?title=Bendigo_railway_line *Contributors*: 712M, Alex1991, Amalas, AxSkov, Axpde, Betacommand, Bryan Derksen, Crusoe8181, Dan027, Garyvines, Grimee, Jeancey, Longhair, Merveism, MoondyneAWB, Peter Horn, Rebecca, Reinthal, Rick Doodle, Rom rulz424, Samantha of Cardyke, Sameboat, Scottius11, Sheepunderscore, Signalhead, Somebody in the WWW, Supt. of Printing, Wongm, Woohookitty, 12 anonymous edits

Outtrim_railway_station *Source*: http://en.wikipedia.org/w/index.php?title=Outtrim_railway_station *Contributors*: Amalas, Billingd, Crusoe8181, Dan027, Longhair, Rich Farmbrough, Wongm

Mildura_railway_line *Source*: http://en.wikipedia.org/w/index.php?title=Mildura_railway_line *Contributors*: AaronRichard, Alex1991, Bleakcomb, Brenont, Commissioner Geoff, Coyets, Crusoe8181, Dan027, Jafeluv, Longhair, Mattinbgn, Peter Horn, Rebecca, ScottDavis, Transaus, WikiSandbox1, Wongm, 12 anonymous edits

North_East_railway_line *Source*: http://en.wikipedia.org/w/index.php?title=North_East_railway_line *Contributors*: Alex1991, Amalas, AxSkov, Axpde, Betacommand, Bidgee, Blarneytherinosaur, Commissioner Geoff, Coyets, Crusoe8181, Dan027, Dl2000, Forthevline, Garyvines, Grahamec, JRG, Jason Recliner, Esq., Kreb, Mattnotley2004, Ninthnotch, Nomadtales, Peter Horn, Philip J. Rayment, Rick Doodle, Rjwilmsi, Rom rulz424, Sameboat, Signalhead, Tabletop, Taitset, Thryduulf, Useddenim, WikiSandbox1, Wongm, Zzrbiker, 32 anonymous edits

Rail_transport_in_Victoria *Source*: http://en.wikipedia.org/w/index.php?title=Rail_transport_in_Victoria *Contributors*: Aitias, Alex1991, Ansett, Axver, Bearcat, Biatch, Bidgee, Bleakcomb, Breno, Brenont, Cassowary, Chowbok, Clarkk, Commissioner Geoff, Crusoe8181, Declan Palmer, Dl2000, Edmund Carew, Elekhh, Emann15, Ewawer, Gaius Cornelius, Garyvines, Gerry Harwood, Grahamec, Grey Shadow, Grimhim, Gunzel, Gwernol, HTML2011, Harryboyles, Hmains, Jamescookmd, Jason Recliner, Esq., John of Reading, Koavf, LJ Holden, Lightmouse, Longhair, Mannafredo, MartinBiely, Mattnotley2004, Melesse, Michael Devore, Michael Johnson, Nathan Johnson, Oxyman42, Ozgunzel, Ozzmosis, Peter Horn, Philip J. Rayment, Plasma east, Polypipe Wrangler, R'n'B, Rebecca, Reinhard Dietrich, Reyk, Richard Felstead, Richsage, Rjwilmsi, Roke, ScottDavis, Somebody in the WWW, Sp33dyphil, Stepshep, Steven J. Anderson, Sturgeonslaw, Supt. of Printing, Syd1435, TPK, Tabletop, Theo Pardilla, Tony1, Transaus, Trevor Tyler, WikHead, WikiDon, Wongm, Zoqaeski, Zzrbiker, 68 anonymous edits

Serviceton_railway_line *Source*: http://en.wikipedia.org/w/index.php?title=Serviceton_railway_line *Contributors*: Adamm, Admella1859, Alex1991, Amalas, Amsterdam biking, AxSkov, Axpde, BTpedia, Coyets, Crusoe8181, Cuddy Wifter, Dan027, Forthevline, Hugo999, Huyie, Jafeluv, Mattinbgn, Mattnotley2004, Ninthnotch, Optimist on the run, Ozzles, Peter E. James, Peter Horn, Picapica, Rebecca, Reinthal, Rich Farmbrough, Rick Doodle, Rjwilmsi, Rom rulz424, Sameboat, Sct72, Shortshadow, Sqlsource68, Topbanana, Wongm, 27 anonymous edits

Orbost_railway_line *Source*: http://en.wikipedia.org/w/index.php?title=Orbost_railway_line *Contributors*: AaronRichard, Axpde, Biscuittin, Black-Velvet, Bleakcomb, Brionnach, Coyets, Crusoe8181, Dan027, Duetral, Jafeluv, Jason Recliner, Esq., Jjron, Jupiter2001, Longhair, Mattinbgn, Mattnotley2004, Nomadtales, Norlag74, Peter Horn, Reinthal, Rick69p, Rom rulz424, Sameboat, Slambo, Soumyasch, Stampjohn, Stepshep, Trainz10, Vmenkov, Wongm, 7 anonymous edits

Portland_railway_line *Source*: http://en.wikipedia.org/w/index.php?title=Portland_railway_line *Contributors*: Coyets, Crusoe8181, Forthevline, John of Reading, Kipperfield, LilHelpa, Peter Horn, Platte, Rjwilmsi, Sameboat, Unusual wally, Wongm, 1 anonymous edits

Port_Fairy_railway_line *Source*: http://en.wikipedia.org/w/index.php?title=Port_Fairy_railway_line *Contributors*: 712M, AxSkov, Axpde, CommonsDelinker, Coyets, Crusoe8181, Eregli bob, Forfucksakepassword, Jafeluv, Mattinbgn, Mattnotley2004, OZJ, Peter Horn, Sameboat, Somebody in the WWW, Wongm, 3 anonymous edits

Sydney–Melbourne_rail_corridor *Source*: http://en.wikipedia.org/w/index.php?title=Sydney%E2%80%93Melbourne_rail_corridor *Contributors*: Arsenikk, Bidgee, Crusoe8181, Darkwind, Frammy7, Giraffedata, Good Olfactory, Hugo999, JRG, Lightmouse, Peter Horn, Reinthal, Sp33dyphil, Thryduulf, Tim Ross, Wongm, 71 anonymous edits

Shepparton_railway_line *Source*: http://en.wikipedia.org/w/index.php?title=Shepparton_railway_line *Contributors*: Alex1991, Amalas, AxSkov, Axpde, Coyets, Crusoe8181, Dan027, Grahamec, It's Monday night, time for Headlines, Jafeluv, Jason Recliner, Esq., Mattinbgn, Peter Horn, Rebecca, Rick Doodle, Rjwilmsi, Rom rulz424, Sameboat, Supt. of Printing, Thryduulf, WikiSandbox1, Wongm, 15 anonymous edits

Melbourne–Adelaide_railway *Source*: http://en.wikipedia.org/w/index.php?title=Melbourne%E2%80%93Adelaide_railway *Contributors*: Arsenikk, Crusoe8181, Cunard, Darkwind, Good Olfactory, Hugo999, JRG, Mattinbgn, Michael Glass, Pdfpdf, Peter Horn, Rich Farmbrough, Shortshadow, Thryduulf, TrackConversion, Ulric1313, Wongm, Wongmihateyou, 5 anonymous edits

Western_standard_gauge_line *Source*: http://en.wikipedia.org/w/index.php?title=Western_standard_gauge_line *Contributors*: Axpde, Coyets, Crusoe8181, Huyie, Pdfpdf, Peter Horn, Reinthal, Rjwilmsi, Sameboat, Shortshadow, Topbanana, WikiSandbox1, Wongm, 7 anonymous edits

Image Sources, Licenses and Contributors

Image:Outtrim-line.png *Source*: http://en.wikipedia.org/w/index.php?title=File:Outtrim-line.png *License*: unknown *Contributors*: Dan027, Rjwilmsi

file:Australia Victoria location map.svg *Source*: http://en.wikipedia.org/w/index.php?title=File:Australia_Victoria_location_map.svg *License*: unknown *Contributors*: Chumwa, NordNordWest, Roke, Uwe Dedering

File:Red pog.svg *Source*: http://en.wikipedia.org/w/index.php?title=File:Red_pog.svg *License*: unknown *Contributors*: Anomie

Image:South-Gippsland-line.png *Source*: http://en.wikipedia.org/w/index.php?title=File:South-Gippsland-line.png *License*: unknown *Contributors*: Beao, Dan027, Rjwilmsi

Image:Melbourne railways map.gif *Source*: http://en.wikipedia.org/w/index.php?title=File:Melbourne_railways_map.gif *License*: unknown *Contributors*: John Shadbolt. Original uploader was JohnnoShadbolt at en.wikipedia

File:BNSF 5350 20040808 Prairie du Chien WI.jpg *Source*: http://en.wikipedia.org/w/index.php?title=File:BNSF_5350_20040808_Prairie_du_Chien_WI.jpg *License*: unknown *Contributors*: User:Slambo

Image:Three rail tracks 350.jpg *Source*: http://en.wikipedia.org/w/index.php?title=File:Three_rail_tracks_350.jpg *License*: unknown *Contributors*: Anomie

File:Brno, Brno Město, historická koňská tramvaj.jpg *Source*: http://en.wikipedia.org/w/index.php?title=File:Brno,_Brno_Město,_historická_koňská_tramvaj.jpg *License*: unknown *Contributors*: User:Aktron/Nápověda

File:5051 Earl Bathurst Cocklewood Harbour.jpg *Source*: http://en.wikipedia.org/w/index.php?title=File:5051_Earl_Bathurst_Cocklewood_Harbour.jpg *License*: unknown *Contributors*: Alex at kms, Duncharris, Geof Sheppard, Nilfanion, RHaworth, Railwayfan2005, Shortfatlad, 1 anonymous edits

File:Shinkansen-type-0.jpg *Source*: http://en.wikipedia.org/w/index.php?title=File:Shinkansen-type-0.jpg *License*: unknown *Contributors*: Cassiopeia sweet, Hohoho, Morio, , 1 anonymous edits

File:CTA tracks.jpg *Source*: http://en.wikipedia.org/w/index.php?title=File:CTA_tracks.jpg *License*: unknown *Contributors*: User:Dschwen

File:Ireland - Dublin - Tram.jpg *Source*: http://en.wikipedia.org/w/index.php?title=File:Ireland_-_Dublin_-_Tram.jpg *License*: unknown *Contributors*: User:CGPGrey

File:2TE10U Russian Locomotive.jpg *Source*: http://en.wikipedia.org/w/index.php?title=File:2TE10U_Russian_Locomotive.jpg *License*: unknown *Contributors*: User:Anthony Ivanoff

File:HŽ 7123 series DMU (06).JPG *Source*: http://en.wikipedia.org/w/index.php?title=File:HŽ_7123_series_DMU_(06).JPG *License*: unknown *Contributors*: User:Orlovic

File:InterCity2 - passenger car interior.jpg *Source*: http://en.wikipedia.org/w/index.php?title=File:InterCity2_-_passenger_car_interior.jpg *License*: unknown *Contributors*: Jonik at en.wikipedia

File:Wagons 550.jpg *Source*: http://en.wikipedia.org/w/index.php?title=File:Wagons_550.jpg *License*: unknown *Contributors*: Original uploader was G-Man at en.wikipedia

Image:Pöörangud Tartu raudteejaamas.jpg *Source*: http://en.wikipedia.org/w/index.php?title=File:Pöörangud_Tartu_raudteejaamas.jpg *License*: unknown *Contributors*: User:DJ Sturm

Image:CTA loop junction.jpg *Source*: http://en.wikipedia.org/w/index.php?title=File:CTA_loop_junction.jpg *License*: unknown *Contributors*: User:Dschwen

File:European railway map.jpg *Source*: http://en.wikipedia.org/w/index.php?title=File:European_railway_map.jpg *License*: unknown *Contributors*: User:PeterEastern

File:Eastbound over SCB.jpg *Source*: http://en.wikipedia.org/w/index.php?title=File:Eastbound_over_SCB.jpg *License*: unknown *Contributors*: Davepape, Gürbetaler, Iain Bell, Ronaldino, Thryduulf, Voyager, 2 anonymous edits

File:HBD DD1.jpg *Source*: http://en.wikipedia.org/w/index.php?title=File:HBD_DD1.jpg *License*: unknown *Contributors*: User:SwalleyD

File:UP 6670.jpg *Source*: http://en.wikipedia.org/w/index.php?title=File:UP_6670.jpg *License*: unknown *Contributors*: terry cantrell

File:Train wreck at Montparnasse 1895.jpg *Source*: http://en.wikipedia.org/w/index.php?title=File:Train_wreck_at_Montparnasse_1895.jpg *License*: unknown *Contributors*: Studio Lévy and Sons (Studio Lévy & fils)

File:DeutscheBahn gobeirne.jpg *Source*: http://en.wikipedia.org/w/index.php?title=File:DeutscheBahn_gobeirne.jpg *License*: unknown *Contributors*: User:gobeirne

File:Stanhope Station Railway Lines.jpg *Source*: http://en.wikipedia.org/w/index.php?title=File:Stanhope_Station_Railway_Lines.jpg *License*: unknown *Contributors*: Tellyaddict

File:JRE-TEC-E5 omiya.JPG *Source*: http://en.wikipedia.org/w/index.php?title=File:JRE-TEC-E5_omiya.JPG *License*: unknown *Contributors*: User:Toshinori baba

Image:Flag of Victoria (Australia).svg *Source*: http://en.wikipedia.org/w/index.php?title=File:Flag_of_Victoria_(Australia).svg *License*: unknown *Contributors*: User:Denelson83, User:Greentubing

Image:Victoria (Australia) coa.gif *Source*: http://en.wikipedia.org/w/index.php?title=File:Victoria_(Australia)_coa.gif *License*: unknown *Contributors*: User:AshGreen

Image:Tigris-Australia location Victoria.svg *Source*: http://en.wikipedia.org/w/index.php?title=File:Tigris-Australia_location_Victoria.svg *License*: unknown *Contributors*: Bidgee, Los688, The Emirr

File:Doudiet Swearing allegiance to the Southern Cross.jpg *Source*: http://en.wikipedia.org/w/index.php?title=File:Doudiet_Swearing_allegiance_to_the_Southern_Cross.jpg *License*: unknown *Contributors*: Billinghurst, Damzow, 2 anonymous edits

File:Victoria Parliament House Melbourne.jpg *Source*: http://en.wikipedia.org/w/index.php?title=File:Victoria_Parliament_House_Melbourne.jpg *License*: unknown *Contributors*: User:Donaldytong

File:Vic legislative council 1878.jpg *Source*: http://en.wikipedia.org/w/index.php?title=File:Vic_legislative_council_1878.jpg *License*: unknown *Contributors*: Chat-rat, Elekhh, Man vyi, Stephen Bain, 1 anonymous edits

File:City Hall, Geelong-Victoria-Australia, 2007.jpg *Source*: http://en.wikipedia.org/w/index.php?title=File:City_Hall,_Geelong-Victoria-Australia,_2007.jpg *License*: unknown *Contributors*: User:Wongm

File:Melbourne skyline.jpg *Source*: http://en.wikipedia.org/w/index.php?title=File:Melbourne_skyline.jpg *License*: unknown *Contributors*: User:Donaldytong

File:Melbourneunisouthlawn.jpg *Source*: http://en.wikipedia.org/w/index.php?title=File:Melbourneunisouthlawn.jpg *License*: unknown *Contributors*: Artemus Jones, Longhair, 2 anonymous edits

File:State Library of Victoria - left side.jpg *Source*: http://en.wikipedia.org/w/index.php?title=File:State_Library_of_Victoria_-_left_side.jpg *License*: unknown *Contributors*: Diliff, Julia W, Ranveig, WingHKLinda, 3 anonymous edits

File:Victorian stand in 1867.jpg *Source*: http://en.wikipedia.org/w/index.php?title=File:Victorian_stand_in_1867.jpg *License*: unknown *Contributors*: User:Rodolph

File:Yallourn-w-power-station-australia.jpg *Source*: http://en.wikipedia.org/w/index.php?title=File:Yallourn-w-power-station-australia.jpg *License*: unknown *Contributors*: User:Wongm

File:Highways of Victoria.PNG *Source*: http://en.wikipedia.org/w/index.php?title=File:Highways_of_Victoria.PNG *License*: unknown *Contributors*: Athaenara

File:C Class Tram, Melbourne - Jan 2008.jpg *Source*: http://en.wikipedia.org/w/index.php?title=File:C_Class_Tram,_Melbourne_-_Jan_2008.jpg *License*: unknown *Contributors*: User:Diliff

Image:Xptnew.jpg *Source*: http://en.wikipedia.org/w/index.php?title=File:Xptnew.jpg *License*: unknown *Contributors*: Original uploader was Vicer at en.wikipedia

Image:X'TrapolisFSS.jpg *Source*: http://en.wikipedia.org/w/index.php?title=File:X'TrapolisFSS.jpg *License*: unknown *Contributors*: Original uploader was Alex1991 at en.wikipedia

Image:Vlocity train at lara victoria.jpg *Source*: http://en.wikipedia.org/w/index.php?title=File:Vlocity_train_at_lara_victoria.jpg *License*: unknown *Contributors*: User:Wongm

File:Victoria_summer.jpg *Source*: http://en.wikipedia.org/w/index.php?title=File:Victoria_summer.jpg *License*: unknown *Contributors*: Original uploader was Artemus Jones at en.wikipedia

File:Victoria_winter.jpg *Source*: http://en.wikipedia.org/w/index.php?title=File:Victoria_winter.jpg *License*: unknown *Contributors*: Original uploader was Artemus Jones at en.wikipedia

File:Victoria_rainfall.jpg *Source*: http://en.wikipedia.org/w/index.php?title=File:Victoria_rainfall.jpg *License*: unknown *Contributors*: Original uploader was Artemus Jones at en.wikipedia

File:TwelveApostles-Gaz.jpg *Source*: http://en.wikipedia.org/w/index.php?title=File:TwelveApostles-Gaz.jpg *License*: unknown *Contributors*: Gaz

File:MCG99.jpg *Source*: http://en.wikipedia.org/w/index.php?title=File:MCG99.jpg *License*: unknown *Contributors*: Original uploader was PaddyBriggs at en.wikipedia

Image:MetlinkSignBase_313.gif *Source*: http://en.wikipedia.org/w/index.php?title=File:MetlinkSignBase_313.gif *License*: unknown *Contributors*: EN WP User Lakeyboy

Image:Danenong station rain.JPG *Source*: http://en.wikipedia.org/w/index.php?title=File:Danenong_station_rain.JPG *License*: unknown *Contributors*: Jamesbehave, Longhair, Shyam

Image:Views from Cricklewood.jpg *Source*: http://en.wikipedia.org/w/index.php?title=File:Views_from_Cricklewood.jpg *License*: unknown *Contributors*: Wongm

Image:Korumburra-layout.GIF *Source*: http://en.wikipedia.org/w/index.php?title=File:Korumburra-layout.GIF *License*: unknown *Contributors*: Dan027

File:Korumburra Commercial Street.jpg *Source*: http://en.wikipedia.org/w/index.php?title=File:Korumburra_Commercial_Street.jpg *License*: unknown *Contributors*: Melburnian

Image:Restored victorian railways train.jpg *Source*: http://en.wikipedia.org/w/index.php?title=File:Restored_victorian_railways_train.jpg *License*: unknown *Contributors*: User:Wongm

File:Spirit of Progress c1938.jpg *Source*: http://en.wikipedia.org/w/index.php?title=File:Spirit_of_Progress_c1938.jpg *License*: unknown *Contributors*: John L. Buckland (d)

Image:67 Spencer Street.jpg *Source*: http://en.wikipedia.org/w/index.php?title=File:67_Spencer_Street.jpg *License*: unknown *Contributors*: Wongm, 1 anonymous edits

Image:Harold Clapp.jpg *Source*: http://en.wikipedia.org/w/index.php?title=File:Harold_Clapp.jpg *License*: unknown *Contributors*: Zzrbiker

Image:S302andB60.jpg *Source*: http://en.wikipedia.org/w/index.php?title=File:S302andB60.jpg *License*: unknown *Contributors*: Photo by Victorian Railways. Original uploader was Zzrbiker at en.wikipedia; a later version with modifications was uploaded by Janke.

Image:Victorian-Railways-wings.png *Source*: http://en.wikipedia.org/w/index.php?title=File:Victorian-Railways-wings.png *License*: unknown *Contributors*: User:Wongm

Image:Vicraillogo.png *Source*: http://en.wikipedia.org/w/index.php?title=File:Vicraillogo.png *License*: unknown *Contributors*: User:Wongm

File:Victorian-rail-map-2007.png *Source*: http://en.wikipedia.org/w/index.php?title=File:Victorian-rail-map-2007.png *License*: unknown *Contributors*: Wongm

Image:Eastern Distributor tunnel.jpg *Source*: http://en.wikipedia.org/w/index.php?title=File:Eastern_Distributor_tunnel.jpg *License*: unknown *Contributors*: user:Nomadtales

Image:Epping station underground platform.JPG *Source*: http://en.wikipedia.org/w/index.php?title=File:Epping_station_underground_platform.JPG *License*: unknown *Contributors*: User:Gareth

File:Green check.svg *Source*: http://en.wikipedia.org/w/index.php?title=File:Green_check.svg *License*: unknown *Contributors*: User:gmaxwell

Image:Passenger-Rail-Australia-Map.png *Source*: http://en.wikipedia.org/w/index.php?title=File:Passenger-Rail-Australia-Map.png *License*: unknown *Contributors*: w:User:MagpieShooterMagpieShooter

Image:Looking east on the trans australia line from cook.jpg *Source*: http://en.wikipedia.org/w/index.php?title=File:Looking_east_on_the_trans_australia_line_from_cook.jpg *License*: unknown *Contributors*: 17177, Fairv8, Infrogmation, Nachoman-au, W.Rebel

Image:Indian Pacific Perth, Western Australia.jpg *Source*: http://en.wikipedia.org/w/index.php?title=File:Indian_Pacific_Perth,_Western_Australia.jpg *License*: unknown *Contributors*: User:Nachoman-au

Image:Sydney Ferries DSC01594.JPG *Source*: http://en.wikipedia.org/w/index.php?title=File:Sydney_Ferries_DSC01594.JPG *License*: unknown *Contributors*: Crusoe8181, Longhair, Yewenyi, 1 anonymous edits

Image:Melbourne--swanston-dock-container-crane.jpg *Source*: http://en.wikipedia.org/w/index.php?title=File:Melbourne--swanston-dock-container-crane.jpg *License*: unknown *Contributors*: User:Wongm

File:Flag of Australia.svg *Source*: http://en.wikipedia.org/w/index.php?title=File:Flag_of_Australia.svg *License*: unknown *Contributors*: Anomie, Mifter

Image:Melbourne Airport 2.jpg *Source*: http://en.wikipedia.org/w/index.php?title=File:Melbourne_Airport_2.jpg *License*: unknown *Contributors*: user:SimonEast

Image:Sydney Airport (2004) By Air.jpg *Source*: http://en.wikipedia.org/w/index.php?title=File:Sydney_Airport_(2004)_By_Air.jpg *License*: unknown *Contributors*: User:Mathieumcguire

Image:PD-icon.svg *Source*: http://en.wikipedia.org/w/index.php?title=File:PD-icon.svg *License*: unknown *Contributors*: Alex.muller, Anomie, Anonymous Dissident, CBM, MBisanz, Quadell, Rocket000, Strangerer, Timotheus Canens, 1 anonymous edits

Image:Vicrailmap-bendigo.png *Source*: http://en.wikipedia.org/w/index.php?title=File:Vicrailmap-bendigo.png *License*: unknown *Contributors*: User:Wongm

File:Big-Hill-tunnel-Bendigo-line.jpg *Source*: http://en.wikipedia.org/w/index.php?title=File:Big-Hill-tunnel-Bendigo-line.jpg *License*: unknown *Contributors*: User:Wongm

File:Castlemaine-station-Melbourne-end.jpg *Source*: http://en.wikipedia.org/w/index.php?title=File:Castlemaine-station-Melbourne-end.jpg *License*: unknown *Contributors*: User:Wongm

Image:Victorian-rail-map-2007.png *Source*: http://en.wikipedia.org/w/index.php?title=File:Victorian-rail-map-2007.png *License*: unknown *Contributors*: Wongm

Image:MaryboroughTrainStation.JPG *Source*: http://en.wikipedia.org/w/index.php?title=File:MaryboroughTrainStation.JPG *License*: unknown *Contributors*: User:Mattinbgn

Image:Vicrailmap-albury.png *Source*: http://en.wikipedia.org/w/index.php?title=File:Vicrailmap-albury.png *License*: unknown *Contributors*: User:Wongm

Image:Spirit of Progress 1953.jpg *Source*: http://en.wikipedia.org/w/index.php?title=File:Spirit_of_Progress_1953.jpg *License*: unknown *Contributors*: Bidgee, Zzrbiker

Image:SoP70s.jpg *Source*: http://en.wikipedia.org/w/index.php?title=File:SoP70s.jpg *License*: unknown *Contributors*: Zzrbiker

Image:Site of Mangalore station.jpg *Source*: http://en.wikipedia.org/w/index.php?title=File:Site_of_Mangalore_station.jpg *License*: unknown *Contributors*: User:Wongm

Image:PN-freight-north-east.jpg *Source*: http://en.wikipedia.org/w/index.php?title=File:PN-freight-north-east.jpg *License*: unknown *Contributors*: User:Wongm

Image:VLine-north-east-line.jpg *Source*: http://en.wikipedia.org/w/index.php?title=File:VLine-north-east-line.jpg *License*: unknown *Contributors*: User:Wongm

Image:LastSteamtoAlbury.jpg *Source*: http://en.wikipedia.org/w/index.php?title=File:LastSteamtoAlbury.jpg *License*: unknown *Contributors*: User:Zzrbiker

Image:AlburyRailwayStationPlatform.jpg *Source*: http://en.wikipedia.org/w/index.php?title=File:AlburyRailwayStationPlatform.jpg *License*: unknown *Contributors*: User:AYArktos

File:Vrmap47vgr.png *Source*: http://en.wikipedia.org/w/index.php?title=File:Vrmap47vgr.png *License*: unknown *Contributors*: Zzrbiker

File:Old Flinders Street Station.jpg *Source*: http://en.wikipedia.org/w/index.php?title=File:Old_Flinders_Street_Station.jpg *License*: unknown *Contributors*: Originally uploaded to by .

File:X32.jpg *Source*: http://en.wikipedia.org/w/index.php?title=File:X32.jpg *License*: unknown *Contributors*: Calliopejen1, Zzrbiker

File:SoP70s.jpg *Source*: http://en.wikipedia.org/w/index.php?title=File:SoP70s.jpg *License*: unknown *Contributors*: Zzrbiker

File:City-loop-exit-spencer-street.jpg *Source*: http://en.wikipedia.org/w/index.php?title=File:City-loop-exit-spencer-street.jpg *License*: unknown *Contributors*: User:Wongm

File:Vline n class train at lara victoria.jpg *Source*: http://en.wikipedia.org/w/index.php?title=File:Vline_n_class_train_at_lara_victoria.jpg *License*: unknown *Contributors*: User:Wongm

File:Mixed-gauge-trackwork-north-geelong.jpg *Source*: http://en.wikipedia.org/w/index.php?title=File:Mixed-gauge-trackwork-north-geelong.jpg *License*: unknown *Contributors*: User:Wongm

File:Mechanical-signalling-north-geelong.jpg *Source*: http://en.wikipedia.org/w/index.php?title=File:Mechanical-signalling-north-geelong.jpg *License*: unknown *Contributors*: User:Wongm

File:Mechanical-interlocking-frame-avenel.jpg *Source*: http://en.wikipedia.org/w/index.php?title=File:Mechanical-interlocking-frame-avenel.jpg *License*: unknown *Contributors*: User:Wongm

File:Led-colour-light-signal.jpg *Source*: http://en.wikipedia.org/w/index.php?title=File:Led-colour-light-signal.jpg *License*: unknown *Contributors*: User:Wongm

File:R707-loco-victorian-railways.jpg *Source*: http://en.wikipedia.org/w/index.php?title=File:R707-loco-victorian-railways.jpg *License*: unknown *Contributors*: User:Wongm

File:Vlocity train at lara victoria.jpg *Source*: http://en.wikipedia.org/w/index.php?title=File:Vlocity_train_at_lara_victoria.jpg *License*: unknown *Contributors*: User:Wongm

File:Vline patronage rail routes chart.jpg *Source*: http://en.wikipedia.org/w/index.php?title=File:Vline_patronage_rail_routes_chart.jpg *License*: unknown *Contributors*: User:Biatch

File:Metropolitan train patronage in melbourne.jpg *Source*: http://en.wikipedia.org/w/index.php?title=File:Metropolitan_train_patronage_in_melbourne.jpg *License*: unknown *Contributors*: User:Biatch

File:Wagons-north-melbourne.jpg *Source*: http://en.wikipedia.org/w/index.php?title=File:Wagons-north-melbourne.jpg *License*: unknown *Contributors*: User:Wongm

File:Pacific national cement train at geelong.jpg *Source*: http://en.wikipedia.org/w/index.php?title=File:Pacific_national_cement_train_at_geelong.jpg *License*: unknown *Contributors*: User:Wongm

File:Restored victorian railways train.jpg *Source*: http://en.wikipedia.org/w/index.php?title=File:Restored_victorian_railways_train.jpg *License*: unknown *Contributors*: User:Wongm

File:Overburden-dredge-yallourn.jpg *Source*: http://en.wikipedia.org/w/index.php?title=File:Overburden-dredge-yallourn.jpg *License*: unknown *Contributors*: Crusoe8181, Wongm

Image:Vicrailmap-ballarat.png *Source*: http://en.wikipedia.org/w/index.php?title=File:Vicrailmap-ballarat.png *License*: unknown *Contributors*: User:Wongm

Image:Serviceton Railway station.jpg *Source*: http://en.wikipedia.org/w/index.php?title=File:Serviceton_Railway_station.jpg *License*: unknown *Contributors*: Scott Davis

Image:VLine VLocity Warrenheip.jpg *Source*: http://en.wikipedia.org/w/index.php?title=File:VLine_VLocity_Warrenheip.jpg *License*: unknown *Contributors*: User:Wongm

Image:Ballarat-station-clocktower.jpg *Source*: http://en.wikipedia.org/w/index.php?title=File:Ballarat-station-clocktower.jpg *License*: unknown *Contributors*: User:Wongm

Image:Vicrailmap-gippsland.png *Source*: http://en.wikipedia.org/w/index.php?title=File:Vicrailmap-gippsland.png *License*: unknown *Contributors*: User:Wongm

File:Line towards Orbost.jpg *Source*: http://en.wikipedia.org/w/index.php?title=File:Line_towards_Orbost.jpg *License*: unknown *Contributors*: User:Wongm

File:Gippsland-line-at-Hernes-Oak.jpg *Source*: http://en.wikipedia.org/w/index.php?title=File:Gippsland-line-at-Hernes-Oak.jpg *License*: unknown *Contributors*: User:Wongm

File:Gippsland line crossed by ICR.jpg *Source*: http://en.wikipedia.org/w/index.php?title=File:Gippsland_line_crossed_by_ICR.jpg *License*: unknown *Contributors*: User:Wongm

File:Walhalla Goldfields Railway.jpg *Source*: http://en.wikipedia.org/w/index.php?title=File:Walhalla_Goldfields_Railway.jpg *License*: unknown *Contributors*: Duncharris, FlickreviewR

File:Bunyip station substation and overhead.jpg *Source*: http://en.wikipedia.org/w/index.php?title=File:Bunyip_station_substation_and_overhead.jpg *License*: unknown *Contributors*: User:Wongm

File:VM 0259 Stratford - Avon River Railway Bridge.jpg *Source*: http://en.wikipedia.org/w/index.php?title=File:VM_0259_Stratford_-_Avon_River_Railway_Bridge.jpg *License*: unknown *Contributors*: User:Vmenkov

Image:Vicrailmap-geelong.png *Source*: http://en.wikipedia.org/w/index.php?title=File:Vicrailmap-geelong.png *License*: unknown *Contributors*: User:Wongm

File:Vlocity train at little river victoria.jpg *Source*: http://en.wikipedia.org/w/index.php?title=File:Vlocity_train_at_little_river_victoria.jpg *License*: unknown *Contributors*: User:Wongm

File:Breakwater-bridge-geelong.jpg *Source*: http://en.wikipedia.org/w/index.php?title=File:Breakwater-bridge-geelong.jpg *License*: unknown *Contributors*: User:Wongm

File:Colac1.jpg *Source*: http://en.wikipedia.org/w/index.php?title=File:Colac1.jpg *License*: unknown *Contributors*: Alex1991, Berichard, Mattinbgn, Melesse, Rebecca

File:Xptnew.jpg *Source*: http://en.wikipedia.org/w/index.php?title=File:Xptnew.jpg *License*: unknown *Contributors*: Original uploader was Vicer at en.wikipedia

MIX
Papier aus verantwortungsvollen Quellen
Paper from responsible sources
FSC® C105338

Printed by Books on Demand GmbH, Norderstedt / Germany